Communications in Computer and Information Science 2068

Rationale

The CCIS series is devoted to the publication of proceedings of computer science conferences. Its aim is to efficiently disseminate original research results in informatics in printed and electronic form. While the focus is on publication of peer-reviewed full papers presenting mature work, inclusion of reviewed short papers reporting on work in progress is welcome, too. Besides globally relevant meetings with internationally representative program committees guaranteeing a strict peer-reviewing and paper selection process, conferences run by societies or of high regional or national relevance are also considered for publication.

Topics

The topical scope of CCIS spans the entire spectrum of informatics ranging from foundational topics in the theory of computing to information and communications science and technology and a broad variety of interdisciplinary application fields.

Information for Volume Editors and Authors

Publication in CCIS is free of charge. No royalties are paid, however, we offer registered conference participants temporary free access to the online version of the conference proceedings on SpringerLink (http://link.springer.com) by means of an http referrer from the conference website and/or a number of complimentary printed copies, as specified in the official acceptance email of the event.

CCIS proceedings can be published in time for distribution at conferences or as post-proceedings, and delivered in the form of printed books and/or electronically as USBs and/or e-content licenses for accessing proceedings at SpringerLink. Furthermore, CCIS proceedings are included in the CCIS electronic book series hosted in the SpringerLink digital library at http://link.springer.com/bookseries/7899. Conferences publishing in CCIS are allowed to use Online Conference Service (OCS) for managing the whole proceedings lifecycle (from submission and reviewing to preparing for publication) free of charge.

Publication process

The language of publication is exclusively English. Authors publishing in CCIS have to sign the Springer CCIS copyright transfer form, however, they are free to use their material published in CCIS for substantially changed, more elaborate subsequent publications elsewhere. For the preparation of the camera-ready papers/files, authors have to strictly adhere to the Springer CCIS Authors' Instructions and are strongly encouraged to use the CCIS LaTeX style files or templates.

Abstracting/Indexing

CCIS is abstracted/indexed in DBLP, Google Scholar, EI-Compendex, Mathematical Reviews, SCImago, Scopus. CCIS volumes are also submitted for the inclusion in ISI Proceedings.

How to start

To start the evaluation of your proposal for inclusion in the CCIS series, please send an e-mail to ccis@springer.com.

Taye Girma Debelee · Achim Ibenthal ·
Friedhelm Schwenker ·
Yehualashet Megersa Ayano
Editors

Pan-African Conference on Artificial Intelligence

Second Conference, PanAfriCon AI 2023
Addis Ababa, Ethiopia, October 5–6, 2023
Revised Selected Papers, Part I

 Springer

Editors
Taye Girma Debelee (iD)
Ethiopian Artificial Intelligence Institute
Addis Ababa, Ethiopia

Friedhelm Schwenker (iD)
Universität Ulm
Ulm, Germany

Achim Ibenthal (iD)
HAWK University of Applied Sciences
and Arts
Göttingen, Germany

Yehualashet Megersa Ayano (iD)
Ethiopian Artificial Intelligence Institute
Addis Ababa, Ethiopia

ISSN 1865-0929   ISSN 1865-0937 (electronic)
Communications in Computer and Information Science
ISBN 978-3-031-57623-2   ISBN 978-3-031-57624-9 (eBook)
https://doi.org/10.1007/978-3-031-57624-9

This Springer imprint is published by the registered company Springer Nature Switzerland AG
The registered company address is: Gewerbestrasse 11, 6330 Cham, Switzerland

Paper in this product is recyclable.

Preface

This edition presents the proceedings of the *Pan-African Conference on Artificial Intelligence 2023 (PanAfriCon AI 2023)*. Starting in 2022, this annual conference focuses on African AI developments. The high demand for such a platform can be seen from a tripling of contributions just within one year. At the same time AI is developing new worldwide trends at an accelerating pace. Examples are generative AI and many professional applications in medical AI, agriculture, autonomous maneuvering, financial technologies, cyber security, office applications, and many more. Related to the African continent, more and more countries are developing AI strategies. It is the set goal of this conference to be an arena for the exchange of best practices and the establishment of joint Pan-African efforts to provide solutions for Africa's key twenty-first century challenges in the social, economic, and ecologic domains.

PanAfriCon AI 2023 aimed at bringing together AI researchers, computational scientists, engineers, entrepreneurs, and decision-makers from academia, industry, and government institutions to discuss the latest trends, opportunities, and challenges of the application of AI in different sectors of the continent. During the conference, attendees were able to exchange the latest information on techniques and workflows used in artificial intelligence in a variety of research fields.

After issuing the call for proposals for the conference in May 2023, 134 contributions were received by August 15, 2023. Based on a single-blind review by 2 reviewers, 71 of the 134 contributions were accepted for presentation and for the further publication process. Following an update considering reviewer comments and Springer CCIS series requirements, these contributions were single-blind reviewed by 3 peer reviewers per paper, out of which 26 submissions were finally accepted for publication in 2 volumes, Springer CCIS 2068 and 2069. The first volume covers medical AI, natural language processing, and text and speech processing, the second AI in finance and cyber security, autonomous vehicles, AI ethics, and life sciences.

Due to logistical considerations, the conference was held completely virtually, organized by the headquarters of the Ethiopian Artificial Intelligence Institute EAII in Addis Ababa. The conference was opened by the director of the EAIC, Worku Gachena. In a keynote speech, Achim Ibenthal, HAWK University of Applied Sciences and Arts, Germany, talked on the elements of a Pan-African AI strategy, spanning a wide range from African history to value chain aspects, opportunities and threats of AI technology. Taye Girma, deputy general director of the research and development cluster of the EAII concluded the conference opening with a session briefing. Four parallel sessions were conducted on October 5 and 6, covering AI in health, AI in services, AI in cyber security, and natural language processing.

This conference would not have been possible without the help of many people and organizations. First of all, we are grateful to all the authors who submitted their contributions. We thank the members of the program committee and the peer reviewers for performing the task of selecting contributions for conference presentations and proceedings with all due diligence.

Finally we hope that readers may enjoy the selection of papers and get inspired by these excellent contributions.

February 2024

Taye Girma Debelee
Achim Ibenthal
Friedhelm Schwenker
Yehualashet Megersa Ayano

Address of the Director General EAII

Worku Gachena Negera
Director General
Ethiopian Artificial Intelligence Institute (EAII)
Addis Ababa
Ethiopia

It is with immense pride and a profound sense of purpose that we present this compilation of papers from the second PanAfriCon AI conference, hosted by our institution, the Ethiopian Artificial Intelligence Institute. Our second conference, conducted virtually in October 2023, brought together a diverse group of thinkers, innovators, and practitioners from across the globe, united in their quest to expand the horizons of artificial intelligence (AI). The thematic areas for this year's proceedings are both a reflection of our current priorities and a forecast of the trajectory AI is poised to take. Natural language processing (NLP) stands at the forefront, exemplifying our commitment to breaking down barriers in communication and making technology accessible to diverse linguistic cultures. The papers in this domain showcase advancements that are not only technical marvels but also bridges to inclusivity. AI in medicine highlights the strides we are making in personalized healthcare, predictive diagnostics, and treatment plans tailored by intelligent systems that learn and adapt. These papers underscore the life-saving potential of AI, offering insights into a future where medicine is intimately informed by machine learning algorithms. The intersection of AI and finance is dissected, revealing how artificial intelligence is reshaping everything from everyday banking to complex investment strategies. The collection presents a visionary perspective on how AI is becoming indispensable in navigating the complexities of the financial world. AI's role in cybersecurity is increasingly critical, and the contributions here detail innovative approaches to thwarting digital threats. These papers are a testimony to AI's evolving capability to act as a guardian in the digital realm, where security is paramount. Autonomous vehicles, an exhilarating field, promise to redefine mobility. The research presented provides a glimpse into the sophisticated AI systems that pilot this transformative technology, ensuring safety and efficiency. Lastly, ethical AI, perhaps the most crucial of all themes, binds the others. The discourse on ethics in AI is not just about guiding principles for AI development but also about the very fabric of the society we aspire to build with these tools. These proceedings are more than just a collection of academic papers; they are a beacon of hope and a map to a future where AI serves humanity, amplifies our potential, and addresses the challenges that span our complex world. As we stand on the cusp of an AI-augmented age, the Ethiopian AI Institute is proud to have been the convener of such a monumental exchange of knowledge. It is my hope that the insights within these pages will inspire and challenge the reader to engage with AI in ways that are profound, positive, and transformative.

Organization

Ethiopian Artificial Intelligence Institute, Addis Ababa, Ethiopia

General Chairs

Taye Girma Debelee	Ethiopian Artificial Intelligence Institute, Ethiopia
Achim Ibenthal	HAWK Univ. of Applied Sciences & Arts, Germany
Friedhelm Schwenker	Ulm University, Germany
Yehualashet Megersa Ayano	Ethiopian Artificial Intelligence Institute, Ethiopia

Program Committee

Fitsum Assamnew	Addis Ababa Institute of Technology, Ethiopia
Abeba Birhane	Mozilla Foundation, San Francisco, Trinity College, Dublin, Ireland
Bisrat Derebessa	Addis Ababa Institute of Technology, Ethiopia
Biniam Gebru	North Carolina A&T State University, USA
Taye Girma Debelee	Ethiopian Artificial Intelligence Institute, Ethiopia
Beakal Gizachew	Addis Ababa Institute of Technology, Ethiopia
Achim Ibenthal	HAWK Univ. of Applied Sciences & Arts, Germany
Worku Jiffara	Adama Science & Technology University, Ethiopia
Solomon Kassa	1888EC, Ethiopia
Yehualashet Megersa Ayano	Ethiopian Artificial Intelligence Institute, Ethiopia
Thomas Meyer	University of Cape Town, South Africa
Samuel Rahimeto	Ethiopian Artificial Intelligence Institute, Ethiopia
Friedhelm Schwenker	Ulm University, Germany
Bruce Watson	Stellenbosch University, South Africa
Teklu Urgessa	Adama Science & Technology University, Ethiopia

Peer Reviewers

The editors and organizing committee sincerely thank the following peer reviewers:

Mesfin Abebe	Adama Science & Technology University, Ethiopia
Allan Anzagira	JPMorgan Chase & Co., USA
Natnael Argaw	Addis Ababa Institute of Technology, Ethiopia
Fitsum Assamnew	Addis Ababa Institute of Technology, Ethiopia
Clayton Baker	University of Cape Town, South Africa
Yohannes Bekele	North Carolina A&T State University, USA
Bisrat Bekele Ergecho	Addis Ababa Institute of Technology, Ethiopia
Sinshaw Bekele Habte	HAWK Univ. of Applied Sciences & Arts, Germany
Victoria Chama	University of Cape Town, South Africa
Bisrat Derebessa	Addis Ababa Institute of Technology, Ethiopia
Biniam Gebru	North Carolina A&T State University, USA
Fraol Gelana	Ethiopian Artificial Intelligence Institute, Ethiopia
Frances Gilles-Webber	University of Cape Town, South Africa
Taye Girma Debelee	Ethiopian Artificial Intelligence Institute, Ethiopia
Beakal Gizachew	Addis Ababa Institute of Technology, Ethiopia
Robbel Habtamu	Addis Ababa Institute of Technology, Ethiopia
Sintayehu Hirphasa	Adama Science & Technology University, Ethiopia
Achim Ibenthal	HAWK Univ. of Applied Sciences & Arts, Germany
Worku Jiffara	Adama Science & Technology University, Ethiopia
Gopi Krishina	Adama Science & Technology University, Ethiopia
Amanuel Kumsa	Ethiopian Artificial Intelligence Institute, Ethiopia
Louise Leenen	University of the Western Cape, South Africa
Surafel Lemma	Addis Ababa Institute of Technology, Ethiopia
Adane Leta	University of Gonder, Ethiopia
Yehualashet Megersa Ayano	Ethiopian Artificial Intelligence Institute, Ethiopia
Tilahun Melak	Adama Science & Technology University, Ethiopia
Chala Merga	Addis Ababa Institute of Technology, Ethiopia
Fitsum Mesfine	Ethiopian Artificial Intelligence Institute, Ethiopia
Thomas Meyer	University of Cape Town, South Africa
Daniel Moges	Ethiopian Artificial Intelligence Institute, Ethiopia
Sudhir Kumar Mohapatra	Sri Sri University, India
Henock Mulugeta	Addis Ababa Institute of Technology, Ethiopia
Abdul-Rauf Nuhu	North Carolina A&T State University, USA

Srinivas Nune	Adama Science & Technology University, Ethiopia
Samuel Rahimeto	Ethiopian Artificial Intelligence Institute, Ethiopia
Friedhelm Schwenker	Ulm University, Germany
Bahiru Shifaw	Adama Science & Technology University, Ethiopia
Ram Sewak Singh	Adama Science & Technology University, Ethiopia
Solomon Teferra	Addis Ababa University, Ethiopia
Menore Tekeba	Addis Ababa Institute of Technology, Ethiopia
Natnael Tilahun	Addis Ababa Science and Technology University, Ethiopia
Tullu Tilahun	Addis Ababa Science and Technology University, Ethiopia
Rosa Tsegaye	Ethiopian Artificial Intelligence Institute, Ethiopia
Amin Tuni	Arsi University, Ethiopia
Teklu Urgessa	Adama Science & Technology University, Ethiopia
Steve Wang	University of Cape Town, South Africa
Bruce Watson	Stellenbosch University, South Africa
Degaga Wolde	Ethiopian Artificial Intelligence Institute, Ethiopia
Leul Wuletaw	University of Michigan, USA
Martha Yifiru	Addis Ababa University, Ethiopia
Azmeraw Yotorawi	Addis Ababa Institute of Technology, Ethiopia

Submission Platform

Friedhelm Schwenker	Ulm University, Germany
Yehualashet Megersa Ayano	Ethiopian Artificial Intelligence Institute, Ethiopia

Collaborating Partner

HAWK University of Applied Sciences and Arts, Göttingen, Germany

Contents – Part I

Contents – Part II

AI Ethics and Life Sciences

Medical AI

Machine Learning Based Stroke Segmentation and Classification from CT-Scan: A Survey

Elbetel Taye Zewde[1]([✉]) [ID], Mersibon Melese Motuma[2] [ID],
Yehualashet Megersa Ayano[1] [ID], Taye Girma Debelee[1,2] [ID],
and Degaga Wolde Feyisa[1] [ID]

[1] Ethiopian Artificial Intelligence Institute, 40782 Addis Ababa, Ethiopia
elbeteltaye70@gmail.com
[2] Department of Computer Engineering, Addis Ababa Science and Technology
University, 120611 Addis Ababa, Ethiopia

Abstract. Brain stroke is a life-threatening condition that requires early
diagnosis to reduce permanent disability and death. The Computed
Tomography (CT) scan is used as a gold standard technique to diagnose brain stroke. However, immediate and accurate interpretation of
the images is challenging, even for skilled neuroradiologists. Therefore,
researchers have focused on developing machine learning (ML) and deep
learning (DL) based systems that are used to detect brain stroke from
CT scan images. The main aim of this study is to review the state-of-
the-art approaches that are used to perform segmentation and classification tasks, the efficiency of existing ML techniques in stroke diagnosis,
the availability of public brain stroke CT scan image datasets, noises
that affect brain CT scan images and denoising techniques, and limitations and challenges of ML techniques in segmentation and classification of brain stroke. A total of 33 papers were identified using inclusion
and exclusion criteria from the results of 7 databases (Science-Direct,
MDPI, Google Scholar, IEEE, Wiley online library, Springer Link, and
Dove Press) from 2018 to June 2023. Where most of the studies utilized
DL-based segmentation and classification techniques. Among the various segmentation techniques used, U-net and U-net-based models are
the dominantly used techniques that exhibit superior performance over
other segmentation models. In contrast, the modified 3D U-Net with the
integration of the squeeze and excitation block has shown the highest
Dice score. Similarly, Convolutional neural networks (CNN) based architectures were found to be dominantly used for brain stroke image classification. Furthermore, in this review, 5 publicly available brain stroke
CT scan image datasets were found. 3 of them have masks and can be
used to train segmentation models. However, due to the limitation in the
subtypes of the images and the number of data that are available in the
repositories to train ML models, most of the reviewed studies have used
local data repositories.

Keywords: Artificial Intelligence · Machine Learning · Deep
Learning · Segmentation · Classification · Stroke

© The Author(s), under exclusive license to Springer Nature Switzerland AG 2024
T. G. Debelee et al. (Eds.): PanAfriConAI 2023, CCIS 2068, pp. 3–45, 2024.
https://doi.org/10.1007/978-3-031-57624-9_1

1   Introduction

Stroke is a disease that is specifically related to blood vessels in the brain. It is also known as a brain attack [26]. It occurs when some tissues in the brain die, causing rupture or blockage (infarction) of the blood vessels. This hinders the brain tissues from getting the required oxygen and nutrients. As a result, parts of the body that are controlled by the damaged brain area can not function properly [65] which results in loss of balance, coordination, vision, speech problems, weakness or paralysis in any part of the body, and involuntary eye movements [25].

Stroke has two major types. These are hemorrhagic and ischemic stroke [63]. Ischemic stroke is caused by an obstruction or blood clot that blocks blood vessels in the brain and leads to sudden loss of brain function. It accounts for approximately 80% of cases of stroke [16,38]. Whereas, hemorrhagic stroke occurs when a weakened blood vessel ruptures and causes normal blood flow to be interrupted and bleed into the brain. This type of stroke is linked with high mortality and severe morbidity [63,142]. Both stroke types are diagnosed using a neuroimaging technique, especially the CT scan. When diagnosing a stroke, physicians will look for the type and location of the stroke that has affected the patient. As shown in Fig. 1, Ischemic stroke manifests as a dark region (hypo dense) well contrasted against its surrounding [116]. While the hemorrhagic stroke manifests as white on CT scan image [8].

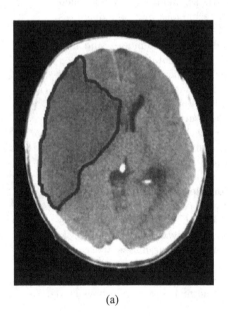
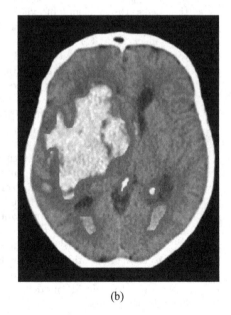

(a)   (b)

Fig. 1. Image in non-contrast CT imaging: (a) ischemic and (b) hemorrhagic stroke.

When a person is suspected of stroke, an early diagnosis is required to deliver the right treatment within the golden time in stroke management within 4.5 h to reduce the risk of permanent disability and/or death [122]. Once a stroke is diagnosed, the initial treatment chosen depends on whether it is ischemic or hemorrhagic type. Therefore, early identification and classification of stroke are essential to decide the type of treatment that needs to be administered.

According to the World Stroke Organization (WSO) [44], in 2019, 63% of the stroke occurred in people younger than 70 years old and 89% of global stroke death and disability combined reside in low to middle-income countries. Stroke is the third leading cause of disability and the second leading cause of death worldwide [65].

1.1 Brian Stroke Image Modalities

Different imaging modalities are used for diagnosing brain stroke. In the health-care sector, 90% of the data exists in the form of images and it is the most important source of evidence for clinical analysis and medical intervention [27,159].

Currently, several imaging techniques are used for diagnosing stroke [104]. Of the existing methods, Computed Tomography Angiography (CTA) is used to detect Large Vessel Occlusions (LVO) of acute ischemic stroke [120]. It is used to determine the narrowing of the carotid arteries [68], but a contrast agent (typically iodine) is required to perform the procedure, and interpretation of the captured image requires a high degree of expertise.

CT Perfusion (CTP) is another technique that is used to capture time-resolved images of blood flow and display formats of perfusion maps including the cerebral blood volume(CBV), Cerebral blood flow(CBF), and mean transit time (MTT) to identify and quantify the volume of infracted core and penumbra. Even though it is easier to interpret than CTA, this method requires longer image acquisition time and is susceptible to motion artifacts. As a result, individuals will be exposed to more radiation doses [131].

The other technique is Non-contrast computed tomography (NCCT) [6,104]. NCCT is considered the primary option for stroke diagnosis. It's known for its quick speed, widespread availability, and cost-efficiency in comparison to other options like magnetic resonance imaging (MRI) and CTP [28]. It is used to diagnose major stroke, identify a contraindication to stroke treatment, and help to identify brain diseases that are similar to a stroke. Even though the CT imaging technique has a low spatial resolution and low sensitivity in the first 3 h after stroke onset for detecting Acute Ischemic Stroke(AIS) and in diagnosing minor strokes, this technique is a fast, widely available, and cost-effective technique among existing ones [66,104,129]. Besides, it is used as the primary imaging technique that is widely used to assess the location and extent of the problem in patients with suspected stroke [104].

At the initial stage, CT scans have a significant role in diagnosing stroke. It is the most common and quick method to diagnose and analyze stroke [16,39, 101]. Diagnosing stroke using CT images has advantages such as fewer spatial constraints and quick shooting time [66,106]. MRI is another imaging modality that has an excellent spatial resolution to detect brain ischemia in transient ischemic attacks or minor ischemic strokes. However, this modality requires a longer time to capture brain images. Besides, the device is not easily available in many healthcare centers [3,104].

1.2 Machine Learning

Machine learning (ML) is a sub-field of Artificial intelligence (AI) that gives computers the ability to develop statistical models and perform a specified task without explicit instructions [82,153]. ML-based systems have several applications in medicine. Along with the growth of technology, it can be implemented to analyze and classify medical images quickly, and accurately and improve clinical outcomes [6,56]. ML is used to offer personalized clinical care for stroke, heart disease detection, and classification [14,129]. This will help to decrease the workload of physicians [148], and reduce the misdiagnosis rate that happens due to the limitations of the existing techniques and personal error. With fast-improving computational power and the availability of enormous amounts of data, Deep learning (DL) has become the widely used ML technique. DL models can learn more sophisticated patterns than conventional ML techniques and greatly simplify the feature engineering process. The ability of neural networks to discover patterns by learning the data and enabling it to identify and classify given data appropriately makes DL important for the medical imaging field [45,110]. So far, many studies have been conducted to explore the use of ML/DL in medical applications for clinical decision-making and monitoring systems [74,148,153,156]. Besides, Lee et al. [74] have mentioned ML techniques' usefulness in screening candidates for therapy among patients with unclear stroke onset time.

2 Related Work

Several review works have been done to examine research on state-of-the-art approaches and challenges toward brain stroke segmentation and classification using ML and DL-based techniques from CT scan images [34,63,95,129,129, 153]. Nur et al. [95] compiled, evaluated, and analyzed the data from relevant research conducted by scholars. According to this review, researchers frequently utilized thresholding, fuzzy C-means, k-means, region growing, and watershed segmentation approaches to separate CT scan images. Furthermore, this review contributes by determining the most effective automated segmentation method for assessing brain stroke.

A review article by Julian et al. [148] provided a detailed review of the application of ML in radiology. The authors thoroughly assessed the availability of medical image datasets to create a powerful ML model. Furthermore, they identified the primary obstacles that hinder the implementation of ML models and provided effective solutions to overcome the challenges.

Inamdar et al. [63] reviewed papers on computer-aided Acute Brain Stroke diagnosis. The authors outlined the state of the art and challenges that computer-aided diagnosis faces in terms of lesion region segmentation and stroke detection. In addition, the authors have mentioned that segmentation models greatly depend on image quality, acquisition, and the reconstruction parameters of the modality [63]. Moreover, Yao *et al.* [153] reviewed the technical characteristics of existing DL applications in neurology. According to the researchers, even if the rapid growth of DL in neuroradiology has shown promising results, the unavailability of implementation methods, the use of internal datasets without external validation, inconsistent assessment metrics, and lack of clinical validation are the main challenges of DL for not being deployed to solve existing problems. Even though the authors provided a detailed discussion on the application of the DL in neurology, they did not further review stroke segmentation and classification. Sirsat *et al.* [129] reviewed ML for brain stroke detection, but they haven't compared state-of-the-art techniques other than ordinary ML algorithms. Additionally, a thorough review [34] was conducted in ischemic image analysis utilizing deep learning techniques to choose suitable site interventions, support clinical practice, and enhance patient outcomes. The first section of this study updated the state-of-the-art deep learning algorithms currently in use and their primary use in acute ischemic stroke imaging, with a focus on investigating the possible roles that multimodal prognostication and stroke diagnosis may play. Finally, the existing problems and prospects were sketched out.

In a review done by Sirsat *et al.* [129], it was reported that the Support Vector Machine (SVM) was found to be the best model in 10 studies for stroke issues. Furthermore, in this study, CT scan images were found to be the most commonly utilized dataset in studies related to stroke. However, the authors mentioned that additional research in the area is required. The contribution and limitation summary of the above-related work is present in Table 1.

3   Method

The main aim of this study is to review papers that focus on ML-based stroke segmentation and/or classification from CT-scan images either using DL or classical ML models and to address the research questions indicated in Table 2. We followed several steps before performing the paper review. These include; determining the keywords used to search for the right papers, selecting potential journals that can be input for the study, determining the inclusion and exclusion criterion, and downloading the search results from the potential databases.

3.1   Research Questions

A skillfully constructed research question can aid in selecting the right study approach and pinpointing the particular paper that needs to be collected and analyzed. As a result, 5 basic questions were raised and addressed in this review. The research questions are indicated in Table 2.

Table 1. Summary of related works.

Author and Year	Contribution	Limitation
Yao et al. [153],2020	Reviewed current algorithms and approaches for neuroimaging technology and Discussed in detail the use of DL in five (5) neuroimaging modalities - MRI, functional MRI, CT, PET, and US	Limited references were reviewed and do not discuss the application of ML for brain stroke segmentation and classification.
Sirsat et al. [129], 2020	Provide a detailed review of machine learning for brain stroke	The comparison did not consider DL techniques other than classical ML.
Manisha et al. [128], 2020	The reviewer identified SVM as the best model compared with current state-of-the-art ML and Suggested further investigation on brain stroke treatment by applying state-of-the-art methods	The authors considered papers published only on the Science Direct database between 2007 and 2019.
Inamdar et al. [63],2021	provided a thorough analysis of the challenges encountered by DL-, ML-, and computer-aided diagnostic (CAD) methods for CT and MRI stroke segmentation and detection	The review only considers the ischemic stroke type and Classification issues were not included in the review.
Cui et al. [34],2022	A comprehensive review was done on ischemic image analysis using deep learning methods, especially in investigating the possible role of multimodal prognostication and stroke diagnosis	The comparison did not consider other stroke types
Julian et al. [148],2020	A Comprehensive review was done on the advantages of AI and ML in the health sector.	The technical feasibility of ML models was not taken into consideration during the discussion

3.2 Search Engine

To ensure the research questions are well addressed, an electronic search was performed according to the Preferred Reporting Items for Systematic Reviews and Meta-Analyses (PRISMA) guidelines for systematic reviews in selected databases. Figure 2 indicates the PRISMA flow diagram of the search process.

The General article searching and selection procedure is indicated in the Algorithm 1.

As indicated in Algorithm 1, First, electronic paper databases were searched and selected. As a result, IEEE Xplore [62], Google Scholar [51], ScienceDirect [123], Springer Link [130], Wiley Online Library [149], Dove Medical Press [40], and MDPI [109] databases were selected. Then, an advanced search strategy was followed by using a set of keywords to get the most out of these electronic search

Table 2. Research questions.

No	Questions
1	Which ML/DL frameworks are commonly used in CT scan image segmentation and classification?
2	Are the existing ML techniques efficient?
3	Are there enough public brain CT-scan image datasets for developing ML models?
4	What are the limitations and challenges of ML and DL-based stroke segmentation and classification?
5	What type of noises affect brain CT scan images? and How can the noises be removed?

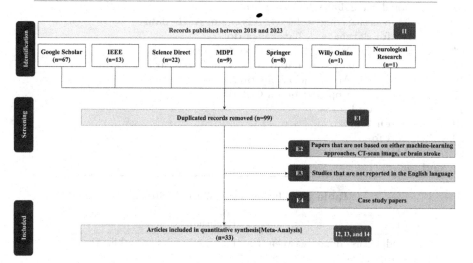

Fig. 2. PRISMA flow diagram for machine learning-based stroke classification and segmentation articles.

databases. Here, the appropriate keywords were selected with their alternatives based on the research questions. Then, Search strings were constructed from keywords by using Boolean AND and OR. The key terms used for the advanced search are shown below:

"Stroke" AND "CT Scan image" AND "Segmentation" OR "Classification" OR "Detection" AND "Machine learning" OR "Deep Learning" OR "Artificial Intelligence" OR "CNN" OR "Hybrid"

Finally, the papers were downloaded and screened based on predefined inclusion and exclusion criteria indicated in Table 3.

Algorithm 1. The pseudocode for paper selection.

1: Databases ← [IEEEXplore, Google Scholar, Springer Link, Dove Medical Press, MDPI, Science Direct, Wiley Online Library]
2: Publication_Year ← [2018 - June 2023]
3: **Initialize Keywords**
4: Disease_Keywords ← [Stroke]
5: Aim_Keywords ← [Classification, Segmentation, Detection, Diagnosis]
6: Method_Keywords ← [Deep Learning, Machine Learning, Artificial Intelligence, ch1CNN, Hybrid]
7: Data_Keywords ← [CT Scan Image]
8: Search_String ← ""
9: **for** disease ∈ Disease_Keywords **do**
10:     **for** aim ∈ Aim_Keywords **do**
11:         **for** data ∈ Data_Keywords **do**
12:             Search_String = disease **AND** aim **AND** data
13:             **for** database ∈ Databases **do**
14:                 **for** Date ∈ Publication_Year **do**
15:                     papers ← databases.search(Search_String)
16:                     Apply Inclusion Criteria
17:                     Apply exclusion Criteria
18:                 **end for**
19:             **end for**
20:         **end for**
21:     **end for**
22: **end for**

Table 3. Inclusion and exclusion criteria.

Inclusion	Exclusion
I1: Studies are done between the years 2018–2023	E1: Duplication of publication
I2: Studies that focus on the segmentation and classification of brain stroke from CT-scan images using Deep learning or any machine learning algorithms	E2: Papers that are not based on either machine-learning approaches, CT-scan image, or brain stroke
I3: Studies that focus on at least one type of stroke	E3: Studies that are not reported in the English language
I4: Studies that are published in Scopus-indexed journals and international conference proceedings	E4: Case study papers

4   ML Models Performance Evaluation Metrics

Reliable model performance assessment is crucial for training and testing ML/DL models [89,90]. There are evaluation metrics and benchmarks used to compare and evaluate ML models. The most commonly used metrics for image segmentation models are the Dice similarity coefficient, Jaccard score, and Hausdorff distance [90]. Additionally, to evaluate the classification model's performance, evaluation metrics like precision, specificity, sensitivity, F1-score, and accuracy of the classifier are commonly used [46]. A confusion matrix is also an evaluation technique that shows the true positive (TP), false positive (FP), true negative (TN), and false negative (FN) results of a classifier. Table 4 shows a confusion matrix [67].

Table 4. Confusion Matrix [46,67].

		Predicted Classes	
		Positive	Negative
Actual Classes	Positive	TP	FP
	Negative	FN	TN

Accuracy. Accuracy (Ac) is the percentage of correctly classified images [50, 60,94]. It is formulated as in the following (Eq. 1).

$$Ac = \frac{TP + TN}{TP + TN + FP + FN} \tag{1}$$

Precision. Precision provides the rate of correctly classified CT scan data with the stroke (TP) to the total CT scan data predicted to have the stroke (TP + FP) [50,94]. Mathematically it is formulated as indicated in (Eq. 2).

$$Pr = \frac{TP}{TP + FP} \tag{2}$$

Recall. Recall (Rc) shows the ratio of correctly classified stroke patients (TP) divided by the total number of patients who have a stroke as shown in (Eq. 3). The logic behind the Recall is how many patients have been classified as having a stroke. Briefly, it is a measure of how correctly the classifier predicted the true positive value [42,50].

$$Rc = \frac{TP}{TP + FN} \tag{3}$$

F-Score. F-score (F1) is the harmonic mean of the ratio of true positive values (recall) and precision. It measures how well the classifier is performing and is often used to compare classifiers. The F1 score specifies the balance between precision and recalls [42,50]. F1-score is calculated as given in (Eq. 4).

$$F1 = 2\frac{Pr \cdot Rc}{Pr + Rc} \tag{4}$$

Dice Similarity Coefficient. The Dice similarity Coefficient is used to compare the likeness of two regions or to measure the overlapping of two regions [97,125]. Given two sets of pixels A and B, the similarity coefficient is defined as shown in (Eq. 5).

$$DCS(A, B) = \frac{2 \, A \cap B|}{|A| + |B|} \tag{5}$$

Where A is the actual value and B is the predicted value.

4.1   Area Under the ROC Curve (AUC)

AUC provides an aggregate measure of performance across all possible classification thresholds as indicated in Fig. 3. It measures the entire two-dimensional area under the curve.

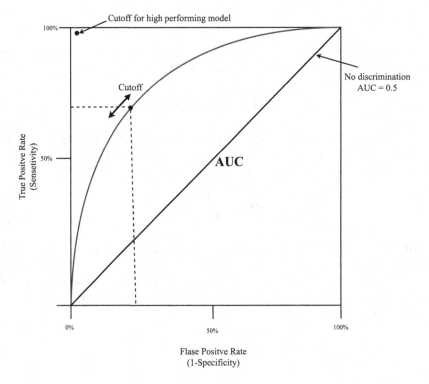

Fig. 3. ROC-curves-and-area-under-curve-AUC [119].

A number between 0.0 and 1.0 represents a binary classification model's ability to separate positive classes from negative classes. The closer the AUC is to 1.0, the better the model's ability to separate classes from each other [113].

Intersection-Over-Union (Jaccard Index). One of the most commonly used metrics in semantic segmentation is Intersection-Over-Union (IoU), also known as the Jaccard Index [22]. It is a very straightforward metric that's extremely effective. The IoU is calculated as the average over all pixels, resulting in an IoU value ranging from 0 to 1. Mathematically, IoU is defined as shown in (Eq. 6).

$$IoU = \frac{|T \cap P|}{|T \cup P|} \tag{6}$$

where T and P correspond to the true label image and prediction of the output image, respectively.

Hausdorff Distance. The Hausdorff distance serves as a widely used metric that measures dissimilarity between sets of points or image segmentation [136] and determines the distance between two sets of points [15]. The average Hausdorff distance between two finite point sets X and Y is defined as in (Eq. 7).

$$d_{\text{AHU}}(X,Y) = \frac{\frac{1}{X}\sum_{x \in X, y \in Y} \min(d(X,Y)) + \frac{1}{Y}\sum_{x \in X, y \in Y} \min(d(X,Y))}{2} \tag{7}$$

The direct average Hausdorff distance from point set X to Y is given by the sum of all minimum distances from all points set X to Y divided by the number of points in X. The average Hausdorff distance can be calculated as the mean of the directed average Hausdorff distance from X to Y and the directed average Hausdorff distance from Y to X.

5   Brain Stroke Image Databases

Over the last few decades, a lot of databases/datasets including Brain Stroke CT scan image datasets were published in different publically available repositories for public use. The available public brain stroke CT scan images are present in either NIFTI file, DICOM format, or JPEG and PNG file formats. Table 5 shows a list of publicly available datasets for brain stroke CT scan images.

6   Artifacts in Brain CT Scan Images and Denoising Techniques

6.1   Noises in Brain CT Scan Images

Artifacts are image distortions or abnormalities that can arise from various sources. These noises or artifacts have the potential to hinder and affect medical images, to the point where they may become diagnostically unreliable [83]. In brain CT scan images, several common artifacts can occur and affect the images [23]. These include; Motion artifacts that occur due to patient movement during the scan [85]. It has the potential to cause a blurring or streaking effect on the

Table 5. Brain CT scan image public datasets.

Dataset	Quantity	Disease class	Segmentation Mask
Murtadha D, *et al.* [57], 2019	82 scans	intraventricular, intraparenchymal, subarachnoid, epidural and subdural	318 slices
Chilamkurthy *et al.* [33], 2018	491 scans	intracranial hemorrhage (ICH) and its types, intraventricular (IVH), subdural (SDH), intraparenchymal (IPH), subarachnoid (SAH), and extradural (EDH) hemorrhages	–
RSNA [121],2019	874,035 slices	epidural, intraparenchymal, intraventricular, subarachnoid, subdural hemorrhage	–
ISLES [1], 2018	156 slices	ischemic stroke	94 slices
MICCAI [80],2021	397 scans	acute ischemic stroke	10972 slices

images. Motion artifacts can be caused when patients are restless, coughing, or due to inadequate immobilization. The other is Ring artifacts, which are circular or ring-shaped noises that occur due to imperfections in the CT scanner [23,85]. These artifacts can lead to concentric rings of increased or decreased attenuation in the images.

Beam hardening and scatter are also artifacts that affect brain CT scan images. Both artifacts produce dark streaks between high-attenuating objects, such as metal, bone, iodinated contrast, or barium. They occur mostly when there is a metal implant in the scanned area [23]. Furthermore, these metal implants can also cause streaking or starburst-like artifacts and Poisson noise due to their high density and the scattering of X-rays. This effect is also known as metal artifacts [23].

The other is partial volume artifacts. It arises when a structure, such as a small lesion or a blood vessel, is only partially included within a voxel (3D pixel) during image reconstruction. This artifact causes blurring or reduced visibility of the structures, making it difficult to accurately assess their shape or size [85].

Furthermore, because of an electrical signal interruption [49], and the limitation of sensors at image acquisition [37] a CT scan image can be affected by Gaussian noise.

Noise level can be calculated by using standard deviation(SD) fluctuations in CT scans [36,83], and Peak Signal-to-noise Ratio (PSNR) [71]. A higher value for SD the greater the noise in CT images [36].

6.2   CT Scan Image Denoising Techniques

To maintain the information, improve the medical image quality, and reduce the noise in CT images, several denoising techniques have been utilized. One of the techniques to reduce the artifacts in brain CT scan images is Iterative reconstruction techniques, such as statistical iterative reconstruction, model-based iterative reconstruction, and adaptive statistical iterative reconstruction techniques. These are mathematical algorithms that are used to iteratively estimate the original image from the acquired data. These techniques can help reduce beam hardening artifacts [85].

The other is the filtered back-propagation technique (FBP). This is a filtering technique that is used before back-propagation to reduce noises and some artifacts. In this technique, smoothing filters or edge-preserving filters are applied to the raw projection data before image reconstruction [23]. Besides, ring artifacts are removed by calibrating the detector, although occasionally the detector itself needs to be replaced [85].

In addition, after image construction, post-processing techniques such as median filtering, adaptive filtering, gaussian filtering, and non-local means filtering are used to reduce noise and enhance image quality.

7   Brain Stroke Segmentation Methods

Image segmentation is the process of clustering an image into several coherent sub-regions. In medical image diagnosis, image segmentation is used for information extraction [118], image simplification into convenient and easier form for analysis [5], and to improve diagnosis accuracy [41,124]. Despite the obstacles encountered in biomedical image segmentation, such as variations in image contrast, organ types, shapes, and orientation, AI-based segmentation has recently become preeminent [135]. Fully automatic segmentation contributed significantly to helping neuroradiologists achieve fast and accurate interpretation [7]. Radiologists use segmentation techniques to detect cognitive abnormalities or bleeding of the brain [134]. There are three ways of image segmentation [93]. Those are semantic segmentation, instant segmentation, and object localization [93].

Semantic image segmentation is a concept of better explaining the overall context of an image. It is the task of classifying each pixel of an image and it helps to identify where and what the objects in the scene [154]. It can be used for improved radiological diagnostics or image-guided interventions [11]. The purpose of semantic segmentation is to segment the input image according to semantic information and predict the semantic category of each pixel from a given label set [87]. For every pixel of an input image, semantic segmentation provides an acceptable inference by predicting labels [54], and it provides more accurate localization than detection [73].

Instance segmentation contains pixel-level classification in semantic segmentation and object detection [152]. Instance segmentation in 3D images identifies small regions within images, particularly those related to biomedical image analysis [158]. Deep learning models are capable of performing well on two-dimensional instance segmentation, however, there are still difficulties associated with 3D instance segmentation due to a lack of sufficient training data caused by the time-consuming and expensive task of properly labeling the correct regions.

Object localization is another way of image segmentation. It is used to find the bounding box coordinates of an object in a given image [73]. By leveraging expert knowledge, bounding box guided can enhance stroke lesion segmentation [100] and it is a faster process than pixel-level annotation. The next section will discuss significant machine learning and deep learning-based brain stroke segmentation methods.

7.1 Classical Machine Learning Based Brain Stroke Segmentation

Thresholding is the simplest method of segmenting images. Badriyah *et al.* [18] presents a promising technique for segmenting stroke in CT scan images using a binarization method through thresholding. The study focuses on two methods of thresholding. These are global binary thresholding and Otsu thresholding. The outcomes of the experiment revealed considerable advancements in terms of image quality, evaluated through two widely accepted metrics: peak signal-to-noise ratio (PSNR) and mean square error (MSE). Of all the segmentation techniques otsu thresholding proved to be the most effective in accurately identifying stroke objects. Optimal results were obtained by setting the lower threshold to a high value of no more than 170. Although a high level of accuracy can be obtained, the thresholding-based segmentation approach often struggles to accurately determine the threshold value. Moreover, the segmentation accuracy of this approach tends to fluctuate when there are variations in the image-scanning illumination.

Tomasetti *et al.* [139] compared fully automated methods based on machine learning and thresholding approach to segment the hypoperfused regions in patients with ischemic stroke. According to their comparison, machine learning-based methods generate more precise results than thresholding approaches. By using random forest(RF), they achieved dice coefficients of 0.68. Even though machine learning outperforms the threshold approaches, the method performs less when the ischemic core and penumbra region are tiny. Besides, Sumijan *et al.* [134] conducted a hybrid thresholding method in detecting and extracting hemorrhage on the CT scan image. An accuracy of 96.43% has been accomplished in segmenting and extracting the region of the brain affected by hemorrhage from CT scan slices.

Nugroho *et al.* [94] developed a multi-segmentation method for hemorrhagic brain stroke from a brain CT scan. The researchers removed the skull region of the brain for further clarity. By using Otsu, active contour, watershed, and FCM clustering methods, the biggest average accuracy value was achieved by using active contours (99.10%), and then the adaptive Otsu average value of 98.78% and next is fuzzy c mean (FCM) with an average value was 98.77%. Although the model appears to perform well, utilizing the Jaccard Index and IoU as performance metrics would provide concrete evidence of its performance. The Summary of classical ML-based brain stroke segmentation is provided in Table 6.

Table 6. Classical machine learning based brain stroke segmentation summary.

Authors	Model	Performance
Tomasetti *et al.* [139]	Random Forest	DCS = 0.68
Sumijan *et al.* [134]	Hybrid thresholding	Accuracy = 96.43%
Badriyah *et al.* [18]	Binarization method through Thresholding	PSNR = 69%
Nugroho *et al.* [94]	Otsu, active contour, watershed, and FCM clustering methods	Accuracy = 99.10%

7.2 Deep Learning Based Brain Stroke Segmentation Methods

Deep learning models can learn from data, and it is suitable for object recognition and segmentation domains. Even though machine learning-based feature extraction has been dominant for a long time, deep learning-based medical image segmentation has promising results [55,107,145]. It can solve various challenges related to poor image enhancement, low accuracy in image classification, and low segmentation accuracy [88].

7.3 Encoder-Decoder Based Image Segmentation

Encoder-decoder-based image segmentation is a technique that uses a neural network with an encoder-decoder architecture as shown in Fig. 4. The encoder part is used for feature extraction, to extract multi-scale semantic features, whereas the decoder is for recovering feature map resolution [59,150]. For semantic image segmentation, the deep learning encoder-decoder model is commonly used. K Hu *et al.* [59] proposes an encoder-decoder CNN (ED-Net) architecture to comprehensively utilize both high-level and low-level semantic information.

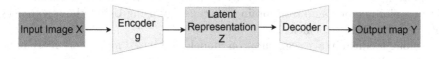

Fig. 4. Simple encoder-decoder model architecture [86].

The encoder performs the encoding operation $z=g(x)$ which is a feature map that is down-sampled by max-pooling layers, and a decoder is used for the up-sampling of features to the original size [76]. The decoding task $Y=f(z)$ is performed by the decoder which predicts Y from the semantic information. Decoder maps the low-resolution encoder feature to full input-resolution feature maps for accurate pixel-wise classification [17].

In another study, Li, Lu *et al.* [76] proposed a deep-learning model(U-net) to segment hemorrhage strokes in CT brain images. The results show that this method can accurately diagnose lesions, making it a useful tool for clinical stroke diagnosis. With the best location accuracy of 0.9859 for detection, 0.8033 dice score, and 0.6919 IoU for segmentation, the proposed model outperformed human experts. The researchers have used manually segmented Brain hemorrhagic CT scans as the ground truth to evaluate U-Net model performance.

Due to the locality of convolution operations, fixed receptive field, and using a single convolutional sequence to extract features at each scale [133], U-net has a limitation in learning global context and long-range spatial relations. To mitigate this problem, Soltanpour *et al.* [131] proposed a new technique called MultiRes U-Net, which uses deep learning and enriched input CTP maps better to identify ischemic stroke lesions from brain CT scans. To analyze multi-resolution spatial information, they have used filters of different sizes (3×3, 5×5, and 7×7). Additionally, the U-net skip connection is replaced by the CNN short cut which is used to combine features from different levels that might result in the semantic gap. An improvement in segmentation task accuracy compared to other previously used methods resulted. This model is an advanced version of the U-Net architecture that includes multi-scale CNN layers and shortcut connections. Enriched input image representations are utilized by the model to enhance accuracy. They achieved an average DSC of 68%. Shaoquan Li *et al.* [77] proposed a multi-scale U-Net deep network model for segmenting acute ischemic stroke from non-enhanced computed tomography. The U-Net architecture consists of convolutional layers that are stacked together, with each layer comprising convolution kernels of uniform size. To achieve better results, instead of relying on a single convolutional kernel, researchers opted for using multiple kernels with varying sizes [77]. The feature fusion of the model is carried out through the kernel of 5×5, 3×3, 1×1 size. By combining these features, one can obtain enhanced features through multiscale convolution, and it improves the classical U-Net angle of view. The researchers achieved the DSC of (0.86 ± 0.04) which is higher than dice based on classical U-net(0.81 ± 0.07) as they compared using the same data sets.

Hssayeni *et al.* [58] developed a deep learning method for segmenting ICH (intracranial hemorrhage) from brain CT scans. By using U-Net the researchers obtained a 0.31 dice coefficient.

Among medical image segmentation models, 2D-based models are more popular than 3D-based segmentation models. But more crucial features can be extracted using 3D than 2D segmentation models [92]. Abramova *et al.* [2] proposed 3D U-net-based hemorrhagic stroke lesion segmentation from a brain CT scan of 76 cases of the clinical data set. The proposed method by the authors utilizes a cutting-edge deep learning approach to segment hemorrhagic stroke lesions in CT scans efficiently. The method involves a robust 3D U-Net architecture enriched with advanced squeeze-and-excitation blocks and restrictive patch sampling, which work seamlessly to deliver accurate results. Their results show promising automated segmentation results, achieving a mean DSC of 0.86 $\pm$

0.074. Even if the 3D network is convenient for the diagnosis of Brain stroke, it is impractical due to computational cost and GPU memory consumption [76].

Omarov *et al.* [97] presents a modified UNet model for brain stroke lesion segmentation on computed tomography images. The proposed model achieves state-of-the-art performance on the ISLES 2018 dataset [1] and outperforms existing methods in accuracy and speed. The authors demonstrate the effectiveness of their approach through experimental results, which show that their model can accurately segment ischemic stroke with a DSC of 58%. They also suggest that their architecture may be extended to other applications, such as ischemia segmentation and CT image identification, by selecting relevant hyperparameters. The proposed methods contribute to the field of medical imaging analysis by proposing an improved deep learning-based method for brain stroke lesion segmentation. The proposed modified UNet architecture incorporates several novel features, such as skip connections, residual blocks, and attention mechanisms, which enhance its performance compared to standard UNet models. This work has potential clinical applications for accurately diagnosing strokes using CT scans or other medical imaging techniques.

To develop medical image segmentation using deep learning, a vast quantity of expert-annotated data is necessary. Platscher *et al.* [107] suggested image-to-image translation models to create magnetic resonance images of brain volumes with and without stroke lesions using semantic segmentation maps to address the data scarcity issue. Moreover, the training of a generative adversarial network to generate synthetic lesion masks is performed. Then, they merge these two elements to construct an extensive synthetic database of stroke images. The U-Net model is employed to evaluate and segment stroke lesions on a clinical test dataset. The best model achieved a Dice score of 72.8%, demonstrating more performance over the others.

7.4   Convolutional Neural Networks Based Brain Stroke Segmentation

CNN is the most extensively utilized deep learning network [10,79], and is used to solve complex problems of image-based pattern recognition [99]. Deep CNN has a better state-of-the-art performance in object segmentation than traditional methods [4]. Barros *et al.* [21] used CNN for the detection and volumetric Segmentation of subarachnoid hemorrhage (SAH) in NCCT. They used 775 NCCTs of brain CT scans, and they obtained a 0.966 intra-class correlation coefficient and an average Dice coefficient of 0.63. According to their report, the CNN-based segmentation is accurate in differentiating SAH from ischemic stroke patients and in segmenting re-bleeds despite the presence of metal artifacts, with an average processing time of 30 s.

Tuladhar *et al.* [140] developed and established a baseline CNN model for automatic NCCT lesion segmentation. They trained and validated a 3D multi-scale CNN model for lesion segmentation using a total of 252 multi-center clinical NCCT datasets obtained from 22 centers, along with their corresponding manual segmentations. The model underwent training using 204 diverse datasets, followed by validation with 48 additional datasets to confirm its accuracy across numerous centers and datasets. Post-processing methods were implemented to improve the CNN-based lesion segmentations. The final CNN model and post-processing method were tested on 39 holdout datasets that are different from the training and validation data sample space. Two or three neuroradiologists segmented each test image. The DSC and predicted lesion volumes were used to evaluate the segmentation. The CNN model achieved a mean DSC score of 0.47 on the validation NCCT datasets. Post-processing improved the DSC to 0.50. On the holdout test set, the CNN model achieved a mean DSC score of 0.42, which was also improved to 0.45 by post-processing. The proposed CNN model can automatically segment ischemic stroke lesions in NCCT datasets.

Shi, T et al. [126] proposed a new DL architecture called cross-modal and cross-attention (C2MA-Net), shown in Fig. 5 The proposed DL network uses cross-modal and cross-attention mechanisms for accurately segmenting AIS lesions from CT perfusion maps. The network has a multipath encoder-decoder architecture and is evaluated using a public dataset of 94 training and 62 test cases. Results show that the C2MA-Net improves recall and F1 scores by 6% and 1%, respectively, compared to other state-of-the-art models.

We have presented the comprehensive summary of strengths and weaknesses of DL models discussed in Sects. 4.2.1 and 4.2.2 in Table 7

Although there have been some advancements in research and architecture for medical image lesion segmentation, further performance improvements are required to effectively apply these techniques to real-world stroke problem-solving [16]. The main reason behind the non-applicability of ML and DL models for real problem-solving is the unavailability of sufficient training datasets, data imbalance, task-dependent training, and low quality of available datasets [30,81,108]. In biomedical image segmentation, UNet-based architectures have outperformed other deep-learning architectures [108]. An additional challenge in improving segmentation performance is that the lesion core of the stroke is small in size compared to overall brain tissues. The lesion core often occupies only around 5% of the total brain tissues [30].

7.5 Semi-supervised Learning Based Brain Stroke Lesion Segmentation

Even if supervised deep learning methods have shown good performance in segmentation tasks, they require a large amount of data that are manually labeled, which is costly and time-consuming to collect [100,145,157]. To overcome this challenge semi-supervised learning-based approaches have been proposed. Semi-supervised learning is a branch of ML that focuses on using both labeled and unlabelled data to learn [143]. If there is a limited number of annotated data, it is good to use semi-supervised based lesion segmentation [35]. A consistent perception generative adversarial network (CPGAN) was proposed by li2017feature et al. [145] for semi-supervised stroke lesion segmentation. A method for semi-supervised stroke lesion segmentation that reduces the dependency on completely labeled data is the proposed CPGAN. It consists of an assistant network to help

Table 7. Summary of deep learning based segmentation of brain stroke model.

Article	Method	Strength	Limitation	Performance
Abramova et al. [2]	3D U-Net	Integration of the squeeze and excitation blocks to the 3D U-Net and addressing the class imbalance because of small lesion volume	3D data labeling and preparation need resources and time	DSC of 0.86 ± 0.074.
Soltanpour et al. [131]	MultiRes U-Net	Extraction of features from a broader range of resolutions and semantic gap reduction	The more filters are added, the higher the time complexity becomes	DSC = 0.68, Jaccard = 0.5713
Tursynova et al. [141]	3D U-Net	More feature is learned in 3D model than 2D	limitation in learning the global context and 3d networks need more resources	-
Li et al. [76]	UNet-based models	Model's prediction for small lesions is very low	The angle of scanning may vary, so it is difficult to have symmetric data all the time	Accuracy = 0.9859, DSC = 0.8033, and IoU = 0.6919
Barros et al. [21]	CNN	High accuracy in segmentation, and classification of stroke	As the lesion size decreases, the model performance also decreases	DSC = 0.63 ± 0.16
Tuladhar et al. [140]	Multiscale 3D CNN	Divers data source provide the model to have general knowledge	Less validation score	DSC = 0.42 − 0.50
Shi, T et al. [126]	C2MA-Net	Improves recall and F2 scores compared to other state-of-the-art	Model requires more time and resources	DSC = 0.4257
Li et al. [77]	Multiscale U-Net	The model performs better than classical U-Net with the same network parameter and improves the classical U-Net angle of view	More than 75% of data is augmented data and the test data was from it	DSC = 0.86 ± 0.04

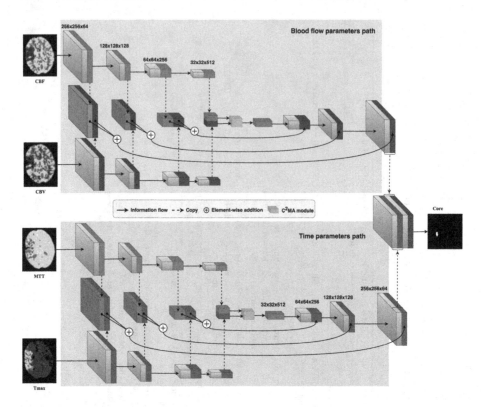

Fig. 5. Cross-model cross-attention network architecture for AIS lesions segmentation, where Cerebral blood flow (CBF), cerebral blood volume (CBV), calculate cerebral blood flow (CBF), mean transit time (MTT), and time-to-maximum(Tmax) [126].

the discriminator acquire meaningful feature representations, a consistent perception technique to improve lesion prediction for unlabeled data, and a similarity connection module (SCM) to collect multi-scale features. The CPGAN-based segmentation obtained a DSC of 0.617 and a Jaccard index of 0.581, according to the authors. The experimental findings show that even with a small amount of labeled data, the proposed CPGAN achieves good segmentation performance.

8   Classification

Image classification is a method of pattern recognition in which images/pixels are classified based on some similarity measures [12]. Brain stroke disease classification from CT scan images is among the areas that have drawn the attention of researchers. It can be done using ML, and DL techniques to help doctors in disease diagnosis or further research. Certain steps are followed for successful brain CT scan image classification. The first step is image pre-processing to improve the quality of the input data and enhance the classification performance of the model. Next is the feature extraction step or extracting effective features from the image. That is, feature extraction helps to obtain visual features from image data. Then the extracted features are used to train the classifier ML model [72]. Classification algorithms use these features and classify the images into defined groups or classes. However, classification algorithms differ in their recognition accuracy, model complexity, computational complexity, memory use, and inference time. Besides, the accuracy of classification models is mainly affected by the qualities of the dataset and the complexity of the analysis problem [98].

8.1   Preprocessing

The quality of the input dataset has a significant impact on how well ML models perform [74]. Alzain *et al.* [9] indicated that the quality of CT scan images can severely be affected by artifacts to the point where they are not useful for diagnosis. Motion artifacts are one of the most frequent artifacts that are usually caused by patient movement; the other is streaking, which occurs near materials such as metal or bone, primarily as a result of beam hardening and scattering. Another distortion that results from several channels gradually diverging from the original measurement is shading. Moreover, distortion and rings can also affect CT Scan images due to errors in the calibration of each detector [144].

Several studies [5, 84, 91, 137] have utilized various preprocessing strategies to avoid artifacts and improve the quality of the CT scan images used for diagnosis. Among the preprocessing methods are: Contrast-limited adaptive histogram equalization which is used to increase the contrast of the images and provide a clearer image [84], and Smoothing with Gaussian blur which is useful for reducing noise. When using Gaussian blur, the kernel size and standard deviation parameters can affect the degree of smoothing and the level of details preserved in the image [47].

Scaling is another technique that serves to regulate the size of the pixels used. Besides, grayscale [137] functions to uniformity of the image's gray degree. Furthermore, to improve the data limitation problem when training machine learning models, Data augmentation techniques are used [91].

The other preprocessing techniques are scaling, which is used to reduce computational cost and shape the data to fit into the classifier model [91], standardization, which is also known as Normalization is used to change the values of pixel intensity using a predefined algorithm to reduce variations from data acquisition [78], Windowing (image window setting) which is used to enhance the contrast of the particular tissue or abnormality type being evaluated. In particular, it provides radiologists with an enhanced view of certain types of cranial abnormalities, such as skull lesions and bone dysplasia.

Preprocessing techniques that help to obtain a clear separation between the brain and skull regions include thresholding, and Brain skull removal which ensures the removal of undesired parts of the image as well as helps to identify the region of interest (ROI) with less competition [5].

8.2   Feature Extraction

Feature extraction is a process of dimension reduction by which an initial set of raw data is reduced to more manageable groups for processing. It aims at extracting the most relevant features that allow one to distinguish among classes [43, 75]. In previous days, medical image features including brain CT scan images were manually extracted by doctors using their professional experience to classify medical images into different classes. But this method is usually difficult (requires experience), and is time-consuming. As a result, it is very prone to instability or no repeatable outcomes [70].

In machine learning and image processing, feature extraction is a process where the features that contribute most to the desired prediction output are automatically or manually selected.

Several feature extraction techniques are applied to brain stroke feature extraction from CT scan images, such as the Regionprops function [5], local binary pattern (LBP) that operates by thresholding neighboring pixel values with the image's center pixel value [53], and the local gradient of the gradient pattern (LG2P) that computes double gradients of local neighborhoods of an original image's center pixel in both x and y directions in the second step [46].

Furthermore, the Structural Co-Occurrence Matrix is also a feature extraction technique that helps to analyze the relationship between low-level structures of two discrete signals in an n-dimensional space. It maps the co-occurrences between the structures of the input signals into a two-dimensional histogram [105, 117].

Similarly, the Gray level co-occurrence matrix (GLCM) is another feature extraction technique that is used to extract texture features of an image by operating according to the second-order statistics in the images [103]. "It describes how frequently two pixels with gray levels G1, and G2 appear in the window separated by a distance d in direction θ. The co-occurrence matrix is a function of two parameters: relative distance measured in pixel numbers (d) and their relative orientation θ" [103].

A convolutional Neural Network has two main parts in its architecture namely the feature extraction part, that is the convolutional layer and pooling layer, and the trainable classifier which is also known as a fully connected neural network is the other technique that can be used for image feature extraction [122].

8.3   Classical-Machine Learning Based Stroke Classification

Several Machine learning algorithms have been used to handle brain stroke classification tasks. For example, the Random forest (RF) technique [111] that fits several decision tree classifiers on various sub-samples of a dataset was used for stroke classification from CT scan images [122, 137]. The RF technique uses averaging to improve the predictive accuracy and to control over-fitting during model training. It has shown high classification accuracy when used to classify stroke disease into two sub-types, namely ischemic stroke and hemorrhagic stroke [137]. In the study, a total of 120 patients' CT scan data was collected and preprocessed in different steps including data conversion, cropping, scaling, grayscaling, and filtering (noise removal)to improve the quality of the data and reduce noise. Then, features including contrast, dissimilarity, homogeneity, correlation, Active Shape Model (ASM), and energy features were extracted using the Gray Level Co-occurrence Matrix (GLCM) method. Finally, eight different models named; K-Nearest Neighbors, Naive Bayes, Logistic Regression, Decision Tree, Random Forest, Multi-layer Perceptron (MLP-NN), Deep Learning, and Support Vector Machine were trained to do the classification task. Finally, the result showed that Random Forest generates the highest level of accuracy (95.97%), along with precision values (94.39%), recall values (96.12%), and f1-Measures (95.39%). However, the authors have claimed that they haven't done any parameter optimization for the classification models and they have used a limited number of data for training the models. As a result, better results could have been achieved.

Furthermore, the RF techniques have shown a promising classification result when used to classify stroke ischemic CT scan images as density changes and no density change images by saragih2020ischemic *et al.* [122]. In this study, 92 images from local repositories, 44 images with density changes, and 48 images with no density change in ischemic stroke patients were used. Then the image features were extracted using the CNN model. Finally, the RF was used as a classifier and showed the best performance (100% accuracy) when 10% of the test dataset was used to evaluate the model and the number of trees used was 100–350 trees. But, an average accuracy of 87% was achieved as the number of testing data increased. From this study, it can be concluded that as the training data set increases and the number of trees increases, the RF models' classification accuracy will increase.

Another machine-learning method used to carry out classification tasks is the Support Vector Machine (SVM). It selects data points that are closest to the classification data (finds a hyperplane to divide two classes). Following that, it will generate the accuracy value, which is based on the kernel value and the input parameters utilized [20]. According to Jain *et al.* [64], SVM is most helpful in cases involving non-linear separation and requires less memory. The SVM classifier can be used by modifying the cubic (polynomial) kernel and Gaussian kernels. A comparison of the Cubic SVM with Gaussian SVM in the classification of infarction for detecting Ischemic Stroke was done by bagasta2019comparison *et al.* [20]. The data set used in the study consists of 206 samples and 7 feature descriptors, namely: Area (cm2) (The size of the area from the infarction point), Minimum value of infarction, the maximum value of infarction, average value of infarction, Standard error value of infarction, Total amount value of infarction, and Length (cm). The results of these experiments were 91.93% (or 92%) accuracy for the Gaussian SVM (RBF Kernel) method and 82.85% (or 82%) accuracy for the Cubic SVM (Polynomial Kernel degree 3). Where, sigma =0.0001 [20]. The results indicate that the Gaussian kernel outperforms the polynomial kernel.

This is because the Gaussian kernel is more flexible than the polynomial kernel when dealing with complex data sets with non-linear boundaries [155].

In another study [46], 9 different machine learning classifiers including the cubic SVM, coarse Gaussian SVM, linear SVM, fine KNN, MKNN, WKNN, ensemble bagged trees, ensemble boosted trees, and ensemble subspace discriminant were used to classify brain CT scan images into three categories; normal, ischemic, and hemorrhagic. In this study, a new feature extraction method named local gradient of gradient pattern (LG2P) was also presented and compared with seven other different feature extraction methods. The LG2P works by first comparing the mean of whole image intensities to the neighbors of the center pixel and next, analyzing double gradients of local neighborhoods of a center pixel of the original image in the x and y directions [46]. Finally, the results show that LG2P is superior to other descriptors including the other variants of Local Binary Pattern (LBP). This includes uniform LBP, rotation invariant LBP, and a combination of the two LBPs. From all 9 classification experiments, using 5-fold and 10-fold cross-validation, the LG2P feature extractor with fine KNN classifier has given the best accuracy (84.44% and 86.11% respectively). This might be due to the limited set of patterns SVM can identify while KNN can find very complex patterns [24]. Besides, the SVM assumes there exists a hyper-plane separating the data points, while KNN attempts to approximate the underlying distribution of the data in a non-parametric way [24].

Similarly, several machine learning techniques were utilized, including the SVM, linear discriminant analysis, minimum learning machine, and multilayer perceptron to classify brain stroke CT scan images into normal, hemorrhagic, and ischemic stroke types [105]. The Structural Co-Occurrence Matrix (SCM) was used with the Fourier transform to extract the most important structural information from the brain CT scan images. The SCM is capable of highlighting the structures of a signal by applying a function k which can change the characteristics of the input signal and generate a new signal k(g). Where the selection of the function K depends on the characteristics of the specific problem. Afterward, the relation of both signals is analyzed in the SCM. In this study, a total of 100 normal, 100 ischemic, and 100 hemorrhagic stroke brain CT scan images were collected. Then the SCM was used to extract the structural features of the images. Finally, the best accuracy (98%) was achieved using the LS-SVM (Least-squares support vector machine) technique with the newly proposed feature extraction technique, which is SCM with Fourier transform. The findings demonstrate that the SCM in the frequency domain can automatically extract the most distinct structural information of strokes, producing good results without the requirement of additional parameters. Table 8 shows the summary papers that used machine learning models to classify brain stroke from CT Scan images.

8.4   Deep Learning-Based Stroke Classification

Studies show that Convolutional neural networks (CNN) have widely been used to diagnose and classify brain stroke from CT scan images [69,84,101,146]. For example; the classification of brain CT scan images as hemorrhagic stroke,

Table 8. Summary of classical machine learning based brain stroke classification.

Article	Objective	Data size	Pre processing	Feature extraction	Classifier	Result
Lori et al. [122]	Ischemic stroke classification into density aud no density images	92 patient data	gray scaling and image resize	CNN	Random forest	Acc = 87%
Tessy et al. [137]	Ischemic and hemorrhagic stroke	226 ischemic and 7 hemorrhagic images	Cropping, Scaling, Greyscale, and data augmentation	GLCM	KNN, Naïve Bayes, Logistic Regression, Decision Tree, MLP, SVM, and RF	Acc = 95.97%, Precision = 94.39%, Recall = 96.12%, and F1-score = 95.39 %
Gautam, et al. [46]	Ischemic, hemorrhagic, and Normal	74 patients comprising 300 slices	Image resize	LG2P	SVM, kNN, ensemble	Acc = 86.11%
Bagast et al. [20]	Classification of infraction for Ischemic stroke identification	206 patient data	-	Area, Minimum value, Maximum value, Average, and Standard error	Cubic SVM, and Gaussian SVM	Acc = 92%
Peixoto et al. [105]	Normal, Hemorrhagic and Ischemic stroke	300 images	-	SCM-Fourier	LS SVM(RBF)	Acc = 98%, Precision = 99.1%, Recall = 97%, and F1-Score =98%

ischemic stroke, and normal CT scan images using the CNN model was done by Marbun *et al.* [84]. The authors used 45 digital public dataset images [112], that is, 10 Ischaemic Stroke images, 10 Hemorrhagic Stroke images, 10 Normal brain images for the Training Dataset, and 5 Hemorrhagic Stroke images, 5 Ischemic Stroke images and 5 Normal brain images for the Testing dataset. All images used were preprocessed using Grayscaling, Scaling, Contrast Limited Adaptive Histogram Equalization, and data augmentation techniques. The thresholding technique, having a threshold value of 170 was implemented to segment the object from the background. The images were then classified using Convolutional Neural Network with a learning rate of 0.2, and a hidden node of 30 for 1000 epochs. Finally, the model has achieved 90% classification accuracy. However, in this research, a limited number of data were used to train the model which limited the model's performance.

Besides, a new CNN architecture called OzNet was introduced as a 2D image classifier [101, 102]. The Oznet is designed with an architecture composed of 34 layers and seven blocks. Each block consists of a convolutional layer, a max pooling layer, an activation function (Rectified linear unit (ReLU)), and a batch normalization layer. Besides, it consists of two fully connected layers Hybrid with various machine learning algorithms for binary classification (stroke and normal class) of brain stroke CT images [101]. The OzNet was coupled with a Decision Tree (DT), k-nearest neighbors (kNN), linear discriminant analysis (LDA), Naïve Bayes (NB), support vector machines (SVM), and a minimum redundancy maximum relevance (mRMR) method (which tends to select the feature that has maximum relevance with respect to the target variable and minimum redundancy with respect to the features that have been selected at previous iterations [114]). A total of 1900 CT images were used for the two classes, and 10-fold cross-validation was utilized to get reliable results while training these algorithms. Additionally, 4096 relevant features were extracted from OzNet's fully connected layer, and the mRMR approach was used to reduce the features' size from 4096 to 250. In the end, the hybrid OzNet-mRMR-NB was shown to give the greatest accuracy of 98.42% and an AUC of 99% to detect stroke from brain CT images. The authors used these machine-learning techniques to classify essential aspects.

Furthermore, using non-linear characteristics in conjunction with a probabilistic neural network classifier, brain CT scan images were classified as normal and hemorrhagic [115]. A total of 1603 non-contrast head CT axial images of 48 patients were used, of which 784 were normal and 819 were ICH. These images were first pre-processed. A mask was initially created for the images and contrast-limited adaptive histogram equalization (CLAHE) was applied to enhance the contrast and intensity of the images. And, a threshold technique was used to extract the brain region. Lastly, seven non-linear features were extracted including; Max. entropy, Yag. entropy, Kap. entropy, Ren. entropy, Shan. entropy, Log entropy, and Vaj. entropy. Finally, using the probabilistic neural network which is a multiclass classifier that uses a kernel discriminant analysis algorithm to classify the new test data into one of the various classes [132], a maximum of 97.37% accuracy was achieved.

Additionally, a 2D CNN classifier and two sequence models comprise the AI method that was introduced [146] for the automatic detection of acute ICH and its subtypes from non-contrast head CT scans. The sequence models learn the correlation across image slices automatically, helping to process large 3D images using 2D CNN models. The 2D CNN was used as a feature extractor, and the model was trained using 2D slices of CT scan image to generate an initial estimation of the presence of one of the ICH (Intracranial Hemorrhage) subtypes (epidural hemorrhage (EDH), subdural hemorrhage (SDH), intraparenchymal hemorrhage (IPH), and subarachnoid hemorrhage (SAH). The sequence model takes the feature outputs from the first-stage classifier and utilizes a bi-directional RNN (recurrent neural network) with the GRU (Gated Recurrent Unit) to produce a refined prediction of the ICH subtypes for every slice. In this study, The RNN model is a regulator that ensures spatially coherent estimation of ICH occurrences while also accounting for 3D context information. Then, a new model was developed to determine the optimal way to combine the predictions from several different current models using another RNN model with the GRU unit. Consequently, the input for the second RNN model (Sequence Model 2) was the combination of the prediction outputs from the first classifier and Sequence Model 1 for all relevant slices of a 3D scan. Following training, Sequence Model 2 outputs each slice of an input head CT with the final prediction of ICH subtypes. Furthermore, slice thickness data was added as an extra input feature to the Sequence Model 2 to allow the model to automatically accommodate variations in slice thickness. As a result, with over 25000 CT scan images from the RSNA-2019 brain CT hemorrhage challenge dataset [121], the model achieved AUCs of 0.984(EDH), 0.988(ICH), 0.996 (IVH),0.992(IPH), 0.983 (SDH), and 0.985(SAH) respectively.

Similarly, the classification of the 5 subtypes of ICH was done using a DL technique for automatic identification and classification of intracranial hemorrhage (ICH) from brain CT images [69]. In this study, the authors first utilized the windowing technique to get three different types of images: bone window, brain window, and subdural window, and performed data augmentation and image resizing in the preprocessing step. Finally, 4,516,842 brain CT images were used to train the CNN-LSTM model. The Xception model was used as a backbone having 36 convolutional layers forming the feature extraction base for the Network. Afterward, 64 nodes and 32-time steps of the LSTM layer were connected to the Xception model. The performance evaluation was done using 727,392 brain CT images, and a resultant 0.07528 weighted multi-label logarithmic loss was achieved, which is approximately equivalent to the classification accuracy of 92 to 93%. The proposed approach can help radiologists read head CT scans and improve the precision of ICH detection and categorization. Likewise, using a total of 752,799 CT scan images from the RSNA Kaggle dataset [121], a new combination of the ResNet model with the EfficientDet model was trained to classify ICH [32]. This is to take advantage of skip connections from the ResNet architecture and use a novel computing unit called the weighted bidirectional feature pyramid network (BiFPN)as the base of the model and a compound scaling method for scaling up several features of the model (resolution, width, and depth), which leads to a new family of object detectors called the EfficientDet family. Furthermore, the Grad-CAM technique was used for a visual explanation of the detected area. Finally, an accuracy of 92.7% and 97.8% ROC-AUC was achieved. In addition, Tharek et al. [138] used the CNN model to detect ICH in head CT scan images. 200 public data was collected and the algorithm model was evaluated and has achieved a sensitivity of 95.9%. However, some cases showed false-positive results due to the presence of calcification in the brain parenchyma.

Additionally, using non-contrast computed tomography (NCCT) images, a deep learning-based automatic detection (DLAD) method was employed to assess the presence of acute ischemic stroke [29]. A 2.5-dimensional neural network architecture, comprising a 2D slice encoder, a slice feature aggregator, and a fully connected classifier, served as the basis for the design of the DLAD algorithm. The feature aggregator, a 1-layer 3D CNN that aggregates the features across multiple slices, received the concatenated feature maps of all the slices. This shows the 3D image's features. The classification result was generated by feeding these features into a two-layer, fully connected classifier. Finally, the DLAD and the ASPECTS readings on non-contrast CT (NCCT) were compared as performed by physicians and were found to have a sensitivity of 65%, specificity of 82%, and accuracy of 80%. The DLAD algorithm was found to be more beneficial for sub-specialists with less experience in reading ASPECTS, allowing quick and accurate assessment of the ASPECTS and timely treatment of acute ischemic stroke.

In another study, a Brain Hemorrhage Classification based on a Neural Network (BHCNet) was introduced to detect brain hemorrhage from CT scan images [91]. The proposed experiment was done using deep-learning classification models including CNN, hybrid CNN + LSTM, and CNN + GRU layers. Data augmentation techniques and image resizing were applied as image pre-processing techniques to increase the quality and quantity of brain CT scan images. Results show that the model achieved 95% accuracy, 90.90% precision, 100% sensitivity, 90% specificity, 95.23% F1-score, and 0.94 area under the curve (AUC).

Furthermore, an adaptive transfer learning technique was used in the classification of ischemic stroke from CT scan images into intact, old infarction, and early ischemic signs [66]. The method optimizes the transfer learning modules to overcome the problems of neural network layer freezing, fine-tuning depth determination strategy selection, and transfer learning velocity. Here, a total of 16,376 2D images from 356 patients were used: 13,064 of the images were used for training, and 3,312 images were used for testing. Normalization, windowing, and image resizing were used in the preprocessing step. Additionally, CNN was combined with many consolidated machine learning techniques, including Bayesian classifiers, Multilayer Perceptron, KNN, Random Forest, and Support Vector Machines, under the Transfer Learning concept. In this study [39], several

CNN models were used including LeNet-5, AlexNet, VGG, Inception, ResNet, Xception, MobileNet, and Nasnet. A total of 420 brain CT scan images-140 healthy and 280 with stroke (140 hemorrhagic and 140 ischemic)-were first collected. Then the CNN models were used as feature extractors, while the machine learning models were used as classifiers. Ultimately, the outcome has shown that, when combined with the majority of the tested classifiers, CNN achieved 100% accuracy, f1 score, recall, and precision. Because it gains the advantage of both models that are used at a time. Similarly, saragih2020ischemic et al. [122] used this technique to classify stroke ischemic CT scan images as density changes and no density change images. The data used were 92 images from local repositories consisting of 48 images with no density changes and 44 images with density change in ischemic stroke patients. In this study, CNN was used for feature extraction, and Random Forest was used as a classifier. As a result, the best performance with 100 percent accuracy was achieved when 10% of the test dataset was used and the number of trees used was 100–350 trees. However, an average accuracy of 87% was achieved as the number of testing data increased.

Not only this, studies have shown that deep learning models' performance highly depends on the hyperparameters selected when training the model [19]. As a result, random search optimization and Bayesian techniques were used to select the best combinations of the hyper-parameters for deep learning with the Multilayer Perceptron (MLP) architecture. In the study by Tessy *et al.* [19], first, the CT scan images were analyzed with grayscale, scaling, thresholding, smoothing, and morphological operations (erosion and dilation). Next, features were selected by using the GLCM. Afterward, the multilayer perceptron deep learning architecture was used to perform the classification task. Lastly, random search and Bayesian techniques were used to set the right hyperparameters for the model. From the results, the Random Search had the best accuracy. In contrast, the optimization time of Bayesian optimization was excellent.

9   Classification and Segmentation

Effective pattern analysis typically involves a preliminary stage where features are extracted or selected to enhance the classification, prediction, or clustering stages. This enables the data to be represented or discriminated in an optimized manner, ability, or better representation of the information. This criterion is necessary because, without first extracting or choosing the relevant features, the raw data are complicated and challenging to handle [48]. Segmentation is a valuable technique that can improve classification accuracy by extracting essential features from an image. By isolating specific areas or regions of an image, segmentation helps highlight details that might have been overlooked. This can ultimately lead to more accurate classifications and more effective analysis of visual data. If you are looking to enhance the accuracy of your image classification models, consider incorporating segmentation into your workflow. However, it is different from feature selection, which better impacts classification. It is effective against noise and loss of information. When we perform segmentation

Table 9. Summary of classical deep learning based brain stroke classification.

Article	Classes	Number of data	Pre-processing	Method	Result
Marbun et al. [84]	Normal, Ischemic, and hemorrhagic	45 brain CT scan images	Grayscaling, Histogram Equalization, Augmentation, and thresholding	CNN	Acc = 90%
Ozaltin et al. [102]	Normal and Stroke	1990 brain CT scan images	Image resize	OZNet-mRMR-NB	Acc = 98.42%
Raghavendra et al. [115]	Normal and hemorrhagic	1603 CT scan images	-	probabilistic neural network	Acc = 97.37% and Recall = 96.94
LI et al. [75]	Epidural, subdural, intraparenchymal, and subarachnoid hemorrhage	25000 CT scan images	-	2D-CNN classifier and Sequential models (RNN model with GRU)	average Acc = 98%
Ko et al. [69]	Epidural, subdural, intraparenchymal, and subarachnoid hemorrhage	727,392 head CT image	-	CNN-LSTM	Acc = 92-93%
Tharek et al. [122]	Epidural, subdural, intraparenchymal, and subarachnoid hemorrhage	200 CT image	-	CNN	Acc = 95%, Precision = 93.14%, Recall = 96.94% and F1-score = 95%
Ferre et al. [32]	Normal and Hemorrhagic	752,799 slices from 18,938 patients [121]	-	ResNet-EfficientDet + Grad-CAM	Acc = 92.7% and 97.8% ROC-AUC

before classification, it allows us to tackle the complexity of object detection and classification in a dataset that contains visually similar categories. Moreover, this approach underscores the significance of modeling subclasses. This approach is essential for the accurate identification and classification of objects, especially in complex scenarios where the differences between categories may be subtle. Through this, we can demonstrate the significance of proper segmentation in improving the performance of classification models [13].

Danfeng Guo *et al.* [52] propose a multi-task fully convolutional network, ICHNet, for simultaneous detection, classification, and segmentation of ICH. The research aims to achieve a set of classification tasks and a segmentation task. One of the classification tasks involves detecting intracranial hemorrhage (ICH), which is classified as binary. The binary ICH detection task is aimed at identifying the presence of ICH in each slice (slice level). The objective of the segmentation task is to accurately categorize every pixel in the CT image into three distinct, non-overlapping classes: hemorrhage, normal brain tissue, and non-brain regions, including the skull and areas outside the skull.

For a precise brain stroke diagnosis, combining image segmentation and classification techniques is crucial rather than using them independently. Yongzhao Xu *et al.* [151] proposed classification followed by a segmentation of brain stroke. During the classification phase, the model distinguishes between an Uninjured brain CT scan and one with a Hemorrhagic stroke. The Mask R-CNN network was utilized to execute the segmentation of the stroke region. The model being proposed was able to achieve an accuracy rate of 100% for class classification and segmentation results with an accuracy of 99.73% and a specificity of 99.93%.

10   Discussion

A brain stroke is a potentially fatal medical condition caused by insufficient blood flow to the brain as a result of a blood vessel rupture or blockage (infarction). The affected brain region will not function normally following a stroke [65], and if treatment is delayed, it may result in death. Besides, it is a global burden, particularly for poorer nations (Johnson, 2016; Feigin, 2022). Early identification is therefore essential for more successful treatment [105].

CT-scan is the most common imaging modality used to investigate brain stroke [147]. However, it requires to be interpreted accurately to deliver proper and on-time treatment for patients. While misdiagnosis rate will be high if the diagnosis using the CT scan is after about 8 d of the first onset of brain stroke symptoms [96]. Therefore, to avoid the misdiagnosis rate and to reduce interpretation time computer-aided diagnosis (CAD) techniques were introduced.

Till now, several research papers focused on utilizing ML and DL techniques to diagnose brain stroke disease from different imaging modalities [63,129,153]. This study focused on the general trends, advancements, and achievements in brain stroke segmentation and classification from CT scan images using classical ML and DL techniques during the 2018–2023 period. Of all the research papers, a total of 33 papers that focus on brain CT scan image segmentation, classification, and both segmentation and classification techniques were selected based on the inclusion and exclusion criteria that are listed in Table 3. The bar graph in Fig. 6 shows the number of reviewed papers based on their category.

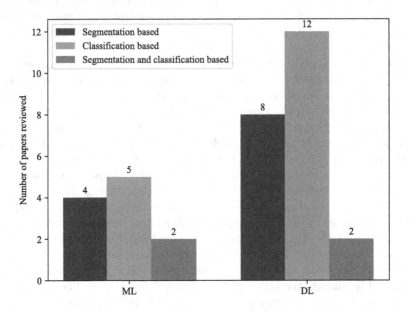

Fig. 6. Bar graph representation for the category of the reviewed papers.

In this review, it can be seen that the studies used diverse datasets, pre-processing and feature extraction techniques, a variety of classification and segmentation models, adaptive techniques, tuning parameters, and different performance metrics. The detailed summary of the reviewed papers is indicated in Table 6, 7, 8, and 9.

Furthermore, through this review, it was found that there are only a limited number of brain stroke CT scan image datasets. In addition, the existing public datasets have limitations in stroke type classification as indicated in Table 5. These datasets focus on either ischemic type [1] or hemorrhagic type and their sub-types only [121], and lack image mask for doing segmentation tasks [33, 121]. As a result, most of the studies used local datasets to do the segmentation and classification tasks.

This review has also indicated the artifacts that frequently affect CT scan images. In most CT scan images, blurring and streaking are the common artifacts that occur due to patients' movements, metallic implants, beam hardening effects, and imperfections in the CT scanner.

Moreover, it was found that different preprocessing and denoising techniques were utilized to improve the quality of the input images. However, each preprocessing and denoising technique has its advantages. For example, the Contrast-limited adaptive histogram equalization technique was used by Marbun *et al.*

[84] to increase the contrast of the images and provide a clearer image. In contrast, Gaussian blur and filtering techniques were used by Tessy *et al.* [137] to have a smooth image and remove noises. Besides, the gray scaling technique was used by saragih2020ischemic *et al.* [122] and Tessy *et al.* [137] to have uniform gray images to make the data ready for training.

Scaling is another technique that was used in most of the reviewed research [46,84,91,101,122,137] to regulate the size of the input images. It has the advantage of reducing computational costs and shaping the data to fit into the models used. Furthermore, the Data augmentation technique was used to improve the data limitation problem when training machine learning models [66,84,91,137]. The other is Windowing (image window setting) which is used to enhance the contrast of the particular tissue or abnormality type being evaluated, was used by Jung *et al.* [66]. In particular, it provides an enhanced view of certain types of cranial abnormalities.

Thresholding was another preprocessing technique that was applied [5]. It was used to remove the skull and ensure the removal of undesired parts of the image as well as help to identify the region of interest (ROI) with less competition.

In addition, with this review, it is revealed that most ML-based brain stroke classification studies [5,46,105,122,137] use different feature extraction techniques to extract important features of the image before doing the classification task. For example, the Regionprops function was used [5] to calculate features of an image such as area, bounding box, centroid, equivalent diameter, eccentricity, extent, convex area, filled area, major axis length, orientation, perimeter, and solidity. It is a built-in function that allows one to measure several features at once. The LBP is another feature extraction technique that was used [46] to extract texture descriptors for brain CT scan images. The LBP technique efficiently captures the local spatial patterns and the grayscale contrast in an image. Similarly, the greyscale co-occurrence matrix is used to extract texture features of an image [103]. Besides, the Structural co-occurrence matrix was the other technique that was used to extract structural features of brain CT scan images [105]. Moreover, CNNs are also used as feature extractors [122]. It can automatically discover and adapt to the most important features of the images, such as edges, textures, shapes, colors, objects, and the spatial and hierarchical structure of the images, by using filters that preserve the local connectivity and context of the pixels. As a result, this will allow CNNs to capture the variability and diversity of the images, and to generalize well to new and unseen data [31]. However, CNNs require a large amount of data to learn adequately.

From the reviewed papers, about 17 studies were found to be focusing on brain stroke classification from CT scan images. Among these papers, 5 were done using classical ML models and 12 were done using DL models. The classical ML models that are used in brain stroke CT scan image classification include the RF technique that gave a promising result using a limited number of data for training the model [122,137], SVM [20,46], KNN [46], naive Bayes and decision tree. From these methods, the SVM and the random forest are found to be frequently used methods. In contrast, CNN is the most commonly used architecture for brain CT scan image classification using deep learning techniques.

Comparing studies or models' classification performance can be challenging due to the variety of performance evaluation metrics used for several tasks, diverse datasets, adaptive techniques, and tuning parameters deployed. But from the studies, it can be concluded that using transfer learning techniques helps to gain the advantages of models that are used to do the classification task at a time [39]. Furthermore, it has been understood that using adaptive transfer learning techniques has a better advantage in reducing the increase in the complexity of transfer learning and operation quantity [66]. Besides, the use of hyperparameter optimization techniques is helpful to get the right value of the hyperparameter and achieve the best result easily [19].

Besides, 12 papers were focused on brain stroke segmentation from CT scan images. In these studies, several encoder-decoder models were used as summarized in Table 7 for image segmentation, and from those U-Net-based model was realized as the most frequently used model for medical image segmentation [141]. It is the standard model for medical image segmentation due to its accuracy with limited labeled training data [41,61,127]. Besides, it is indicated that more crucial features can be extracted using 3D networks than 2D segmentation models [92]. However, it is impractical due to computational cost and memory consumption. As a result, in medical image segmentation models, 2D-based models are more popular than 3D-based segmentation models. Generally, from the reviewed papers it can be concluded that the practical utilization of ML techniques has the potential to improve diagnosis and classification of stroke from CT scan images. It will help to improve the diagnosis and classification accuracy, reduce response time, and reduce the misdiagnosis rate by assisting professionals in the analysis of brain CT-scan images. This review paper provides readers with a comprehensive discussion on the efficiency of existing ML techniques, the availability of datasets, noise in CT scan images, denoising techniques, and limitations and challenges of ML techniques in brain stroke diagnosis. Therefore, researchers can use this as a base to understand the recently existing techniques used in this area and perform further research to develop a real-time applicable machine-learning-based tool to assist physicians.

11 Challenges and Future Directions

To develop a robust ML model used for segmentation and classification, there are several challenges researchers are facing today. Those are:

- There is no sufficient publicly available brain stroke CT scan image.
- The existing public datasets have limited stroke classes and masks for each image in the subclasses.
- Data imbalance
- Computational resource limitation to use 3D models for segmentation and classification.
- Noises that hinder the visibility of lesions affected by stroke.

To achieve an ML model for brain stroke segmentation and classification that can be highly effective, researchers in this area need to collect a considerable amount of valuable data. Proper organization of the classes for classification and acquiring the necessary masks for segmentation are also essential steps. These measures will ensure that the model can be successfully utilized in real-world brain stroke diagnosis. Additionally, it is possible to enhance the performance of models by using semi-supervised ML techniques and data augmentation methods to solve data scarcity.

12 Conclusion

In this study, 33 papers that focused on brain stroke segmentation and classification from CT scan images using classical ML or DL models were reviewed. From these studies, it can be understood that CT scans are the commonly used modality that is used to diagnose brain stroke disease. In most cases, motion artifacts, beam hardening artifacts, ring artifacts, metal artifacts, partial volume artifacts, and beam hardening artifacts from the sinuses are found to be the commonly occurring noise types. Whereas the iterative reconstruction techniques, filtered backpropagation with preprocessing, sinogram restoration techniques, and metal artifact reduction techniques can be applied as pre-processing techniques that can be applied before image construction to remove the artifacts. Besides, as post-processing techniques median filtering, adaptive filtering, and non-local mean filtering can be used to remove the artifacts. Furthermore, from this review, it's important to note that deep learning and machine learning techniques were used and have shown promising results in stroke segmentation and classification from CT scan images. However, comparing all the studies was challenging due to the variety of tasks performed (subtypes considered), the difference in the performance metrics used for different tasks, diverse datasets, adaptive techniques, and tuning parameters deployed. However, it can be concluded that the Unet model architecture is the most frequently utilized technique for brain stroke segmentation from CT scan images. In contrast, CNN is the most commonly used architecture for brain CT scan image classification. Furthermore, to achieve better diagnostic results, both segmentation and classification techniques were used. That is by segmenting the ROI first and then classifying the image classes second.

References

1. Isles challenge 2018 ischemic stroke lesion segmentation. ISLES2018 (2018). http://www.isles-challenge.org/. Accessed 24 Feb 2023
2. Abramova, V., et al.: Hemorrhagic stroke lesion segmentation using a 3d u-net with squeeze-and-excitation blocks. Comput. Med. Imaging Graph. **90**, 101908 (2021)
3. Acharya, U.R., et al.: Automatic detection of ischemic stroke using higher order spectra features in brain MRI images. Cogn. Syst. Res. **58**, 134–142 (2019)

4. Akilan, T., Wu, Q.J., Safaei, A., Huo, J., Yang, Y.: A 3d CNN-LSTM-based image-to-image foreground segmentation. IEEE Trans. Intell. Transp. Syst. **21**(3), 959–971 (2019)

5. Alawad, D.M., Mishra, A., Hoque, M.T.: AIBH: accurate identification of brain hemorrhage using genetic algorithm based feature selection and stacking. Mach. Learn. Knowl. Extract. **2**(2), 56–77 (2020)

6. Alexander, A., Jiang, A., Ferreira, C., Zurkiya, D.: An intelligent future for medical imaging: a market outlook on artificial intelligence for medical imaging. J. Am. Coll. Radiol. **17**(1), 165–170 (2020)

7. Ali, N.H., Abdullah, A.R., Saad, N.M., Muda, A.S., Sutikno, T., Jopri, M.H.: Brain stroke computed tomography images analysis using image processing: A review. Int. J. Artif. Intell. ISSN **2252**(8938), 1049 (2021)

8. Allison, M.: Ct image of hemorrhagic stroke (2019). https://www.accessmedicinenetwork.com/posts/50918-ct-image-of-hemorrhagic-stroke/

9. Alzain, A.F., et al.: Common computed tomography artifact: source and avoidance. Egypt. J. Radiol. Nucl. Med. **52**(1), 151 (2021)

10. Alzubaidi, L., et al.: Review of deep learning: concepts, CNN architectures, challenges, applications, future directions. J. Big Data **8**, 1–74 (2021)

11. Asgari Taghanaki, S., Abhishek, K., Cohen, J.P., Cohen-Adad, J., Hamarneh, G.: Deep semantic segmentation of natural and medical images: a review. Artif. Intell. Rev. **54**, 137–178 (2021)

12. Asokan, A., Anitha, J., Ciobanu, M., Gabor, A., Naaji, A., Hemanth, D.J.: Image processing techniques for analysis of satellite images for historical maps classification-an overview. Appl. Sci. **10**(12), 4207 (2020)

13. Audebert, N., Le Saux, B., Lefèvre, S.: Segment-before-detect: vehicle detection and classification through semantic segmentation of aerial images. Remote Sens. **9**(4), 368 (2017)

14. Ayano, Y.M., Schwenker, F., Dufera, B.D., Debelee, T.G.: Interpretable machine learning techniques in ECG-based heart disease classification: a systematic review. Diagnostics **13**(1), 111 (2023)

15. Aydin, O.U., et al.: On the usage of average hausdorff distance for segmentation performance assessment: hidden error when used for ranking. Eur. Radiol. Exp. **5**, 1–7 (2021)

16. Babutain, K., Hussain, M., Aboalsamh, H., Al-Hameed, M.: Deep learning-enabled detection of acute ischemic stroke using brain computed tomography images. Int. J. Adv. Comput. Sci. Appl. **12**(12), 386–397 (2021)

17. Badrinarayanan, V., Kendall, A., Cipolla, R.: Segnet: a deep convolutional encoder-decoder architecture for image segmentation. IEEE Trans. Pattern Anal. Mach. Intell. **39**(12), 2481–2495 (2017)

18. Badriyah, T., Sakinah, N., Syarif, I., Syarif, D.R.: Segmentation stroke objects based on ct scan image using thresholding method. In: 2019 First International Conference on Smart Technology & Urban Development (STUD), pp. 1–6. IEEE (2019)

19. Badriyah, T., Santoso, D.B., Syarif, I., Syarif, D.R.: Improving stroke diagnosis accuracy using hyperparameter optimized deep learning. Int. J. Adv. Intell. Inf. **5**(3), 256 (2019)

20. Bagasta, A., Rustam, Z., Pandelaki, J., Nugroho, W.: Comparison of cubic SVM with gaussian SVM: classification of infarction for detecting ischemic stroke. IOP Conf. Ser. Mater. Sci. Eng. **546**, 05201 (2019)

21. Barros, R.S., et al.: Automated segmentation of subarachnoid hemorrhages with convolutional neural networks. Inf. Med. Unlocked **19**, 100321 (2020)

22. van Beers, F., Lindström, A., Okafor, E., Wiering, M.A.: Deep neural networks with intersection over union loss for binary image segmentation. In: ICPRAM, pp. 438–445 (2019)
23. Boas, F.E., Fleischmann, D., et al.: Ct artifacts: causes and reduction techniques. Imaging Med. 4(2), 229–240 (2012)
24. Bzdok, D., Krzywinski, M., Altman, N.: Machine learning: supervised methods. Nat. Methods 15(1), 5 (2018)
25. Cao, Z., et al.: Deep learning derived automated aspects on non-contrast ct scans of acute ischemic stroke patients. Technical report, Wiley Online Library (2022)
26. Centers for disease control and prevention (2020). https://www.cdc.gov/stroke/about.htm. Accessed 10 Jan 2023
27. Chan, H.P., Samala, R.K., Hadjiiski, L.M., Zhou, C.: Deep learning in medical image analysis, pp. 3–21 (2020)
28. Chen, W., et al.: Improving the diagnosis of acute ischemic stroke on non-contrast CT using deep learning: a multicenter study. Insights Imaging 13(1), 1–12 (2022)
29. Chiang, P.L., et al.: Deep learning-based automatic detection of aspects in acute ischemic stroke: improving stroke assessment on CT scans. J. Clin. Med. 11(17), 5159 (2022)
30. Clerigues, A., Valverde, S., Bernal, J., Freixenet, J., Oliver, A., Lladó, X.: Acute ischemic stroke lesion core segmentation in CT perfusion images using fully convolutional neural networks. Comput. Biol. Med. 115, 103487 (2019)
31. Convolutional neural network (2017). https://www.engati.com/glossary/convolutional-neural-network. Accessed 30 June 2023
32. Cortés-Ferre, L., Gutiérrez-Naranjo, M.A., Egea-Guerrero, J.J., Pérez-Sánchez, S., Balcerzyk, M.: Deep learning applied to intracranial hemorrhage detection. J. Imaging 9(2), 37 (2023)
33. Cq500 (2018). http://headctstudy.qure.ai/dataset. Accessed 3 Apr 2023
34. Cui, L., et al.: Deep learning in ischemic stroke imaging analysis: a comprehensive review. BioMed Res. Int. 2022 (2022)
35. Cui, W., et al.: Semi-supervised brain lesion segmentation with an adapted mean teacher model. In: Chung, A.C.S., Gee, J.C., Yushkevich, P.A., Bao, S. (eds.) IPMI 2019. LNCS, vol. 11492, pp. 554–565. Springer, Cham (2019). https://doi.org/10.1007/978-3-030-20351-1_43
36. Diwakar, M., Kumar, M.: A review on CT image noise and its denoising. Biomed. Signal Process. Control 42, 73–88 (2018)
37. Dogra, A., Goyal, B., Agrawal, S., Vig, R.: Filtering techniques eliminate gaussian image noise (2023). https://www.vision-systems.com/print/content/14174546
38. Donkor, E.S.: Stroke in the 21st century: a snapshot of the burden, epidemiology, and quality of life. Stroke Res. Treatment (2018). https://doi.org/10.1155/2018/3238165
39. Dourado, C.M., Jr., et al.: Deep learning iot system for online stroke detection in skull computed tomography images. Comput. Netw. 152, 25–39 (2019)
40. Dovepress. https://www.dovepress.com/. Accessed date 23 Nov 2022
41. Du, G., Cao, X., Liang, J., Chen, X., Zhan, Y.: Medical image segmentation based on u-net: a review. J. Imaging Sci. Technol. 64, 1–12 (2020)
42. Faghani, S., et al.: Mitigating bias in radiology machine learning: 3. performance metrics. Radiol. Artif. Intell. 4(5), e220061 (2022)
43. Feature extraction (2020). https://deepai.org/machine-learning-glossary-and-terms/feature-extraction. Accessed 27 Feb 2023
44. Feigin, V.L., et al.: World stroke organization (WSO): global stroke fact sheet 2022. Int. J. Stroke 17(1), 18–29 (2022)

45. Fu, G.S., Levin-Schwartz, Y., Lin, Q.H., Zhang, D.: Machine learning for medical imaging (2019)

46. Gautam, A., Raman, B.: Local gradient of gradient pattern: a robust image descriptor for the classification of brain strokes from computed tomography images. Pattern Anal. Appl. **23**, 797–817 (2020)

47. Gedraite, E.S., Hadad, M.: Investigation on the effect of a gaussian blur in image filtering and segmentation. In: Proceedings ELMAR-2011, pp. 393–396. IEEE (2011)

48. Ghojogh, B., et al.: Feature selection and feature extraction in pattern analysis: a literature review. arXiv preprint arXiv:1905.02845 (2019)

49. Goyal, B., Agrawal, S., Sohi, B.: Noise issues prevailing in various types of medical images. Biomed. Pharmacol. J. **11**(3), 1227 (2018)

50. Grandini, M., Bagli, E., Visani, G.: Metrics for multi-class classification: an overview. arXiv preprint arXiv:2008.05756 (2020)

51. Google scholar. https://scholar.google.com/. Accessed 23 Nov 2022

52. Guo, D., et al.: Simultaneous classification and segmentation of intracranial hemorrhage using a fully convolutional neural network. In: 2020 IEEE 17th International Symposium on Biomedical Imaging (ISBI), pp. 118–121. IEEE (2020)

53. Hadid, A.: The local binary pattern approach and its applications to face analysis. In: 2008 First Workshops on Image Processing Theory, Tools and Applications, pp. 1–9 (2008). https://doi.org/10.1109/IPTA.2008.4743795

54. Hafiz, A.M., Bhat, G.M.: A survey on instance segmentation: state of the art. Int. J. Multimedia Inf. Retr. **9**(3), 171–189 (2020)

55. Hesamian, M.H., Jia, W., He, X., Kennedy, P.: Deep learning techniques for medical image segmentation: achievements and challenges. J. Digit. Imaging **32**, 582–596 (2019)

56. Holzinger, A., Langs, G., Denk, H., Zatloukal, K., Müller, H.: Causability and explainability of artificial intelligence in medicine. Wiley Interdisc. Rev. Data Min. Knowl. Disc. **9**(4), e1312 (2019)

57. Hssayeni, M.: Computed tomography images for intracranial hemorrhage detection and segmentation (2018). https://physionet.org/content/ct-ich/1.0.0/. https://doi.org/10.13026/w8q8-ky94

58. Hssayeni, M.D., Croock, M.S., Salman, A.D., Al-khafaji, H.F., Yahya, Z.A., Ghoraani, B.: Intracranial hemorrhage segmentation using a deep convolutional model. Data **5**(1), 14 (2020)

59. Hu, K., et al.: Automatic segmentation of intracerebral hemorrhage in ct images using encoder-decoder convolutional neural network. Inf. Process. Manag. **57**(6), 102352 (2020)

60. Hussain, E., Hasan, M., Hassan, S.Z., Azmi, T.H., Rahman, M.A., Parvez, M.Z.: Deep learning based binary classification for alzheimer's disease detection using brain mri images. In: 2020 15th IEEE Conference on Industrial Electronics and Applications (ICIEA), pp. 1115–1120. IEEE (2020)

61. Ibtehaz, N., Rahman, M.S.: Multiresunet: rethinking the u-net architecture for multimodal biomedical image segmentation. Neural Netw. **121**, 74–87 (2020)

62. IEEE (1884). https://ieeexplore.ieee.org/Xplore/home.jsp. Accessed 23 Nov 2022

63. Inamdar, M.A., et al.: A review on computer aided diagnosis of acute brain stroke. Sensors **21**(24), 8507 (2021)

64. Jain, U., et al.: Cubic SVM classifier-based feature extraction and emotion detection from speech signals. In: 2018 International Conference on Sensor Networks and Signal Processing (SNSP), pp. 386–391 (2018)

65. Johnson, W., Onuma, O., Owolabi, M., Sachdev, S.: Stroke: a global response is needed. Bull. World Health Organ. **94**(9), 634 (2016)
66. Jung, S.M., Whangbo, T.K.: A deep learning system for diagnosing ischemic stroke by applying adaptive transfer learning. J. Internet Technol. **21**(7), 1957–1968 (2020)
67. Kaya, B., Önal, M.: A CNN transfer learning-based approach for segmentation and classification of brain stroke from noncontrast ct images. Int. J. Imaging Syst. Technol. **33**, 1335–1352 (2023)
68. Kleindorfer, D.O., et al.: 2021 guideline for the prevention of stroke in patients with stroke and transient ischemic attack: a guideline from the american heart association/american stroke association. Stroke **52**(7), e364–e467 (2021)
69. Ko, H., Chung, H., Lee, H., Lee, J.: Feasible study on intracranial hemorrhage detection and classification using a cnn-lstm network. In: 2020 42nd Annual International Conference of the IEEE Engineering in Medicine & Biology Society (EMBC), pp. 1290–1293. IEEE (2020)
70. Krizhevsky, A., Sutskever, I., Hinton, G.E.: Imagenet classification with deep convolutional neural networks. Commun. ACM **60**(6), 84–90 (2017)
71. Kumar, M., Diwakar, M.: A new exponentially directional weighted function based ct image denoising using total variation. J. King Saud Univ.-Comput. Inf. Sci. **31**(1), 113–124 (2019)
72. Lai, Z., Deng, H.: Medical image classification based on deep features extracted by deep model. Comput. Intell. Neurosci. **2018** (2018). https://doi.org/10.1155/2018/2061516
73. Lan, S., et al.: Discobox: weakly supervised instance segmentation and semantic correspondence from box supervision. In: Proceedings of the IEEE/CVF International Conference on Computer Vision, pp. 3406–3416 (2021)
74. Lee, H., et al.: Machine learning approach to identify stroke within 4.5 hours. Stroke **51**(3), 860–866 (2020)
75. Li, J., et al.: Feature selection: a data perspective. ACM Comput. Surv. (CSUR) **50**(6), 1–45 (2017)
76. Li, L., et al.: Deep learning for hemorrhagic lesion detection and segmentation on brain ct images. IEEE J. Biomed. Health Inf. **25**(5), 1646–1659 (2020)
77. Li, S., Zheng, J., Li, D.: Precise segmentation of non-enhanced computed tomography in patients with ischemic stroke based on multi-scale u-net deep network model. Comput. Methods Programs Biomed. **208**, 106278 (2021)
78. Li, X.T., Huang, R.Y.: Standardization of imaging methods for machine learning in neuro-oncology. Neuro-oncol. Adv. **2**(Supplement_4), iv49–iv55 (2020)
79. Li, Z., Liu, F., Yang, W., Peng, S., Zhou, J.: A survey of convolutional neural networks: analysis, applications, and prospects. IEEE Trans. Neural Netw. Learn. Syst. **33**, 6999–7019 (2021)
80. Liang, K., et al.: Symmetry-enhanced attention network for acute ischemic infarct segmentation with non-contrast CT images. In: de Bruijne, M., et al. (eds.) MICCAI 2021. LNCS, vol. 12907, pp. 432–441. Springer, Cham (2021). https://doi.org/10.1007/978-3-030-87234-2_41
81. Liu, X., Song, L., Liu, S., Zhang, Y.: A review of deep-learning-based medical image segmentation methods. Sustainability **13**(3), 1224 (2021)
82. Mahesh, B.: Machine learning algorithms-a review. Int. J. Sci. Res. (IJSR) **9**, 381–386 (2020)
83. Manson, E., Ampoh, V.A., Fiagbedzi, E., Amuasi, J., Flether, J., Schandorf, C.: Image noise in radiography and tomography: causes, effects and reduction techniques. Curr. Trends Clin. Med. Imaging **2**(5), 555620 (2019)

84. Marbun, J., Andayani, U., et al.: Classification of stroke disease using convolutional neural network. In: Journal of Physics: Conference Series, vol. 978, p. 012092. IOP Publishing (2018)

85. Marshall, E.L., Ginat, D.T., Sammet, S.: Computed tomography imaging artifacts in the head and neck region: pitfalls and solutions. Neuroimaging Clin. N. Am. **32**(2), 271–277 (2022)

86. Minaee, S., Boykov, Y.Y., Porikli, F., Plaza, A.J., Kehtarnavaz, N., Terzopoulos, D.: Image segmentation using deep learning: a survey. IEEE Trans. Pattern Anal. Mach. Intell. **44**, 3523–3542 (2021)

87. Mo, Y., Wu, Y., Yang, X., Liu, F., Liao, Y.: Review the state-of-the-art technologies of semantic segmentation based on deep learning. Neurocomputing **493**, 626–646 (2022)

88. Moorthy, J., Gandhi, U.D.: A survey on medical image segmentation based on deep learning techniques. Big Data Cogn. Comput. **6**(4), 117 (2022)

89. Müller, D., Hartmann, D., Meyer, P., Auer, F., Soto-Rey, I., Kramer, F.: Miseval: a metric library for medical image segmentation evaluation. In: Challenges of Trustable AI and Added-Value on Health. Proceedings of MIE (2022)

90. Müller, D., Soto-Rey, I., Kramer, F.: Towards a guideline for evaluation metrics in medical image segmentation. BMC. Res. Notes **15**(1), 1–8 (2022)

91. Mushtaq, M.F., et al.: Bhcnet: neural network-based brain hemorrhage classification using head ct scan. IEEE Access **9**, 113901–113916 (2021)

92. Nodirov, J., Abdusalomov, A.B., Whangbo, T.K.: Attention 3D u-net with multiple skip connections for segmentation of brain tumor images. Sensors **22**(17), 6501 (2022)

93. Novozámský, A., et al.: Automated object labeling for cnn-based image segmentation. In: 2020 IEEE International Conference on Image Processing (ICIP), pp. 2036–2040. IEEE (2020)

94. Nugroho, A.K., Putranto, T.A., Purnama, I.K.E., Purnomo, M.H.: Multi segmentation method for hemorraghic detection. In: 2018 International Conference on Intelligent Autonomous Systems (ICoIAS), pp. 62–66. IEEE (2018)

95. Nur, H.A., Abdul, R.A., Norhashimah, M.S., Ahmad, S.M., Tole, S., Mohd, H.J.: Brain stroke computed tomography images analysis using image processing: a review. IAES Int. J. Artif. Intell. (IJ-AI) **10**(4), 1048–1059 (2021). https://doi.org/10.11591/ijai.v10.i4.pp1048-1059

96. Ogun, S., Oluwole, O., Ogunseyinde, A., Fatade, B., Odusote, K.: Misdiagnosis of stroke-a computerized tomography scan study. West Afr. J. Med. **19**(1), 19–22 (2000)

97. Omarov, B., et al.: Modified unet model for brain stroke lesion segmentation on computed tomography images. Comput. Mater. Continua **71**(3), 4701–4717 (2022)

98. Orenes, Y., Rabasa, A., Rodriguez-Sala, J.J., Sanchez-Soriano, J.: Benchmarking analysis of the accuracy of classification methods related to entropy. Entropy **23**(7), 850 (2021)

99. O'Shea, K., Nash, R.: An introduction to convolutional neural networks. arXiv preprint arXiv:1511.08458 (2015)

100. Ou, Y., et al.: Bbox-guided segmentor: leveraging expert knowledge for accurate stroke lesion segmentation using weakly supervised bounding box prior. Comput. Med. Imaging Graph. **107**, 102236 (2023)

101. Ozaltin, O., Coskun, O., Yeniay, O., Subasi, A.: A deep learning approach for detecting stroke from brain ct images using oznet. Bioengineering **9**(12), 783 (2022)

102. Ozaltin, O., Coskun, O., Yeniay, O., Subasi, A.: Classification of brain hemorrhage computed tomography images using oznet hybrid algorithm. Int. J. Imaging Syst. Technol. **33**(1), 69–91 (2023)

103. Öztürk, Ş, Akdemir, B.: Application of feature extraction and classification methods for histopathological image using glcm, lbp, lbglcm, glrlm and sfta. Procedia Comput. Sci. **132**, 40–46 (2018)

104. Patil, S., Rossi, R., Jabrah, D., Doyle, K.: Detection, diagnosis and treatment of acute ischemic stroke: current and future perspectives. Front. Med. Technol. **4**, 748949 (2022)

105. Peixoto, S.A., Rebouças Filho, P.P.: Neurologist-level classification of stroke using a structural co-occurrence matrix based on the frequency domain. Comput. Electr. Eng. **71**, 398–407 (2018)

106. Phaphuangwittayakul, A., Guo, Y., Ying, F., Dawod, A.Y., Angkurawaranon, S., Angkurawaranon, C.: An optimal deep learning framework for multi-type hemorrhagic lesions detection and quantification in head CT images for traumatic brain injury. Appl. Intell. 1–19 (2022)

107. Platscher, M., Zopes, J., Federau, C.: Image translation for medical image generation: ischemic stroke lesion segmentation. Biomed. Signal Process. Control **72**, 103283 (2022)

108. Prajapati, R., Kwon, G.R.: Sip-unet: sequential inputs parallel unet architecture for segmentation of brain tissues from magnetic resonance images. Mathematics **10**(15), 2755 (2022)

109. Mdpi (1996). https://www.mdpi.com/. Accessed 23 Nov 2022

110. Fuzzy C-Means-Grey Wolf Optimization for classification of Stroke (2021)

111. Sruthi, E.R.: Understand random forest algorithms with examples (2023). https://www.analyticsvidhya.com/blog/2021/06/understanding-random-forest/

112. Radiopaedia (2005). https://radiopaedia.org/. Accessed 24 Mar 2023

113. Classification: Roc curve and auc (2020). https://developers.google.com/machine-learning/crash-course/classification/roc-and-auc. Accessed 24 Feb 2023

114. Radovic, M., Ghalwash, M., Filipovic, N., Obradovic, Z.: Minimum redundancy maximum relevance feature selection approach for temporal gene expression data. BMC Bioinf. **18**(1), 1–14 (2017)

115. Raghavendra, U., et al.: Novel and accurate non-linear index for the automated detection of haemorrhagic brain stroke using ct images. Complex Intell. Syst. **7**, 929–940 (2021)

116. Rajini, N.H., Bhavani, R.: Computer aided detection of ischemic stroke using segmentation and texture features. Measurement **46**(6), 1865–1874 (2013)

117. Ramalho, G.L.B., Ferreira, D.S., Rebouças Filho, P.P., de Medeiros, F.N.S.: Rotation-invariant feature extraction using a structural co-occurrence matrix. Measurement **94**, 406–415 (2016)

118. Ramesh, K., Kumar, G.K., Swapna, K., Datta, D., Rajest, S.S.: A review of medical image segmentation algorithms. EAI Endorsed Trans. Perv. Health Technol. **7**(27), e6–e6 (2021)

119. Roc curve (2020). https://www.researchgate.net/figure/ROC-curves-and-area-under-curve-AUC_fig2_351506473. Accessed 30 June 2023

120. Reidler, P., et al.: Performance of automated attenuation measurements at identifying large vessel occlusion stroke on ct angiography. Clin. Neuroradiol. **31**, 763–772 (2021)

121. RSNA: Rsna intracranial hemorrhage detection (2019). https://www.kaggle.com/competitions/rsna-intracranial-hemorrhage-detection/data

122. Saragih, G.S., Rustam, Z., Aldila, D., Hidayat, R., Yunus, R.E., Pandelaki, J.: Ischemic stroke classification using random forests based on feature extraction of convolutional neural networks. Int. J. Adv. Sci. Eng. Inf. Technol. **10**(5), 2177 (2020)

123. Science direct. https://www.sciencedirect.com/. Accessed 23 Nov 2022

124. Seo, H., et al.: Machine learning techniques for biomedical image segmentation: an overview of technical aspects and introduction to state-of-art applications. Med. Phys. **47**(5), e148–e167 (2020)

125. Shamir, R.R., Duchin, Y., Kim, J., Sapiro, G., Harel, N.: Continuous dice coefficient: a method for evaluating probabilistic segmentations. arXiv preprint arXiv:1906.11031 (2019)

126. Shi, T., Jiang, H., Zheng, B.: C2 ma-net: cross-modal cross-attention network for acute ischemic stroke lesion segmentation based on ct perfusion scans. IEEE Trans. Biomed. Eng. **69**(1), 108–118 (2021)

127. Siddique, N., Paheding, S., Elkin, C.P., Devabhaktuni, V.: U-net and its variants for medical image segmentation: a review of theory and applications. IEEE Access **9**, 82031–82057 (2021)

128. Sirsat, M., Fermé, E., Câmara, J.: Machine learning for brain stroke: a review. J. Stroke Cerebrovasc. Dis. **29** (2020). https://doi.org/10.1016/j.jstrokecerebrovasdis.2020.105162

129. Sirsat, M.S., Fermé, E., Câmara, J.: Machine learning for brain stroke: a review. J. Stroke Cerebrovasc. Dis. **29**(10), 105162 (2020)

130. Springer. https://www.link.springer.com/. Accessed 12 Mar 2022

131. Soltanpour, M., Greiner, R., Boulanger, P., Buck, B.: Improvement of automatic ischemic stroke lesion segmentation in ct perfusion maps using a learned deep neural network. Comput. Biol. Med. **137**, 104849 (2021)

132. Specht, D.F.: Probabilistic neural networks. Neural Netw. **3**(1), 109–118 (1990)

133. Su, R., Zhang, D., Liu, J., Cheng, C.: Msu-net: multi-scale u-net for 2d medical image segmentation. Front. Genet. **12**, 639930 (2021)

134. Sumijan, S., Yuhandri, Y., Boy, W.: Hybrid thresholding method in detection and extraction of brain hemorrhage on the ct-scan image. J. Comput. Sci. Inf. Technol. 7–14 (2021)

135. Suri, J.S., et al.: Unet deep learning architecture for segmentation of vascular and non-vascular images: a microscopic look at unet components buffered with pruning, explainable artificial intelligence, and bias. IEEE Access **11**, 595–645 (2022)

136. Taha, A.A., Hanbury, A.: An efficient algorithm for calculating the exact hausdorff distance. IEEE Trans. Pattern Anal. Mach. Intell. **37**(11), 2153–2163 (2015)

137. Badriyah, T., Sakinah, N., Syarif, I., Syarif, D.R.: Machine Learning Algorithm for Stoke Disease Classification (2020)

138. Tharek, A., Muda, A.S., Hudi, A.B., Hudin, A.B.: Intracranial hemorrhage detection in ct scan using deep learning. Asian J. Med. Technol. **2**(1), 1–18 (2022)

139. Tomasetti, L., Hollesli, L.J., Engan, K., Kurz, K.D., Kurz, M.W., Khanmohammadi, M.: Machine learning algorithms versus thresholding to segment ischemic regions in patients with acute ischemic stroke. IEEE J. Biomed. Health Inf. **26**(2), 660–672 (2021)

140. Tuladhar, A., Schimert, S., Rajashekar, D., Kniep, H.C., Fiehler, J., Forkert, N.D.: Automatic segmentation of stroke lesions in non-contrast computed tomography datasets with convolutional neural networks. IEEE Access **8**, 94871–94879 (2020)

141. Tursynova, A., Omarov, B.: 3D u-net for brain stroke lesion segmentation on isles 2018 dataset. In: 2021 16th International Conference on Electronics Computer and Computation (ICECCO), pp. 1–4. IEEE (2021)
142. Unnithan, A.K.A., Das, J.M.: Hemorrhagic stroke. StatPearls (2022)
143. Van Engelen, J.E., Hoos, H.H.: A survey on semi-supervised learning. Mach. Learn. **109**(2), 373–440 (2020)
144. Veikutis, V., et al.: Artifacts in computer tomography imaging: how it can really affect diagnostic image quality and confuse clinical diagnosis? J. Vibroeng. **17**(2), 995–1003 (2015)
145. Wang, S., Chen, Z., You, S., Wang, B., Shen, Y., Lei, B.: Brain stroke lesion segmentation using consistent perception generative adversarial network. Neural Comput. Appl. **34**(11), 8657–8669 (2022)
146. Wang, X., et al.: A deep learning algorithm for automatic detection and classification of acute intracranial hemorrhages in head ct scans. NeuroImage: Clin. **32**, 102785 (2021)
147. Wardlaw, J.M., Keir, S.L., Dennis, M.S.: The impact of delays in computed tomography of the brain on the accuracy of diagnosis and subsequent management in patients with minor stroke. J. Neurol. Neurosurg. Psychiat. **74**(1), 77–81 (2003). https://doi.org/10.1136/jnnp.74.1.77, https://jnnp.bmj.com/content/74/1/77
148. Wichmann, J.L., Willemink, M.J., De Cecco, C.N.: Artificial intelligence and machine learning in radiology: current state and considerations for routine clinical implementation. Invest. Radiol. **55**(9), 619–627 (2020)
149. Wiley online library. https://www.onlinelibrary.wiley.com/. Accessed 12 Mar 2022
150. Xing, Y., Zhong, L., Zhong, X.: An encoder-decoder network based fcn architecture for semantic segmentation. Wirel. Commun. Mobile Comput. **2020** (2020)
151. Xu, Y., et al.: Deep learning-enhanced internet of medical things to analyze brain ct scans of hemorrhagic stroke patients: a new approach. IEEE Sens. J. **21**(22), 24941–24951 (2020)
152. Yang, S., Wang, X., Yang, Q., Dong, E., Du, S.: Instance segmentation based on improved self-adaptive normalization. Sensors **22**(12), 4396 (2022)
153. Yao, A.D., Cheng, D.L., Pan, I., Kitamura, F.: Deep learning in neuroradiology: a systematic review of current algorithms and approaches for the new wave of imaging technology. Radiol. Artif. Intell. **2**(2) (2020)
154. Yu, H., et al.: Methods and datasets on semantic segmentation: a review. Neurocomputing **304**, 82–103 (2018)
155. Zanaty, E.: Support vector machines (SVMs) versus multilayer perception (MLP) in data classification. Egypt. Inf. J. **13**(3), 177–183 (2012)
156. Zhang, Z., Sejdić, E.: Radiological images and machine learning: trends, perspectives, and prospects. Comput. Biol. Med. **108**, 354–370 (2019)
157. Zhao, B., et al.: Automatic acute ischemic stroke lesion segmentation using semi-supervised learning. arXiv preprint arXiv:1908.03735 (2019)
158. Zhao, Z., Yang, L., Zheng, H., Guldner, I.H., Zhang, S., Chen, D.Z.: Deep learning based instance segmentation in 3D biomedical images using weak annotation. In: Frangi, A.F., Schnabel, J.A., Davatzikos, C., Alberola-López, C., Fichtinger, G. (eds.) MICCAI 2018. LNCS, vol. 11073, pp. 352–360. Springer, Cham (2018). https://doi.org/10.1007/978-3-030-00937-3_41
159. Zhou, S.K., et al.: A review of deep learning in medical imaging: Imaging traits, technology trends, case studies with progress highlights, and future promises. Proc. IEEE **109**(5), 820–838 (2021)

Multitask Deep Convolutional Neural Network with Attention for Pulmonary Tuberculosis Detection and Weak Localization of Pathological Manifestations in Chest X-Ray

Degaga Wolde Feyisa[1]([⊠]) [iD], Yehualashet Megersa Ayano[1] [iD],
Taye Girma Debelee[1,2] [iD], and Samuel Sisay Hailu[3] [iD]

[1] Ethiopian Artificial Intelligence Institute, 40782 Addis Ababa, Ethiopia
degagawolde@gmail.com
[2] Department of Computer Engineering, Addis Ababa Science and Technology University, 120611 Addis Ababa, Ethiopia
[3] Department of Radiology, School of Medicine, Addis Ababa University, 11760 Addis Ababa, Ethiopia

Abstract. Pulmonary tuberculosis (PTB) is a highly fatal bacterial infection that affects the lungs. Chest radiography is a commonly used technique for PTB diagnosis. Interpreting chest X-ray images for features like cavitation, consolidation, and nodules poses challenges due to low contrast between lesions and surrounding tissue, and the complexity of identifying features for intricate disorders. To address these challenges, researchers have proposed using deep learning techniques to detect and mark areas of TB infection in chest X-rays. However, fully supervised semantic segmentation requires massive large pixel-by-pixel labeled images, which is time-consuming, expensive, and subjective. As a result, there is growing interest in weak localization techniques, a method identifying disease pathologies without pixel-level labeling. Hence, this study focuses on developing a deep learning model for weakly supervised segmentation and localization of radiographic manifestations of PTB from chest X-rays (CXR), using commonly used public datasets for TB identification. We proposed multi-scale attention using the DenseNet-121 model as a backbone network. First, a class activation map is calculated at different levels of the backbone network using the last feature map and the global average pooling at each specific level. Finally, the class activation map is combined using a convex combination and passed to the sigmoid functions. This approach is powerful for classifying and localizing disease pathology in CXR. We achieved a localization accuracy of 83% for T (IoU) $= 0.1$ and the classification AUC, accuracy and F_1 score are 98%, 98%, and 97% respectively. This result indicates the model has a promising performance in both the classification and localization of PTB manifestations.

Keywords: Chest X-Ray · Pulmonary Tuberculosis · Weak localization · Weakly Supervised

T. G. Debelee et al. (Eds.): PanAfriConAI 2023, CCIS 2068, pp. 46–59, 2024.
https://doi.org/10.1007/978-3-031-57624-9_2

1   Introduction

PTB is a contagious respiratory infection caused by Mycobacterium tuberculosis. It is spread through the air and can cause symptoms such as cough, chest pain, fatigue, and weight loss, and it can be diagnosed through various tests and treated with antibiotics. If left untreated, it can be life-threatening and may spread to other parts of the body [5]. According to the World Health Organization (WHO), in 2019, around 10 million people had active TB disease, and 1.4 million died from it. The global TB report for 2021 revealed that the target of reducing TB incidence by 20% between 2015 and 2020 was missed, with only an 11% decrease achieved, and the mortality reduction target of 35% was also not met, as death rates only improved by 9.2% [7,8]. This highlights that TB continues to be a leading cause of death.

The diagnosis of active tuberculosis is often based on a symptoms inquiry questionnaire and chest radiography, as these have been found to have a higher sensitivity than other diagnostic approaches, and CXR abnormalities often correspond to symptoms [13,26,34]. The sensitivity and specificity of the symptoms inquiry questionnaire are 77% and 66%, respectively, while they are higher for CXR (86% and 89%, respectively) [12,31]. CXR is considered an effective diagnostic tool for tuberculosis (TB) as it can detect lung abnormalities caused by TB, even in asymptomatic or mild cases. It is widely available, non-invasive, and relatively inexpensive, making it useful in resource-limited settings. However, CXR has limitations, such as lower specificity, which can lead to false positives, so it is often used in conjunction with other diagnostic approaches to confirm TB diagnosis.

The radiographic features of PTB refer to the characteristic patterns that can be seen on chest X-rays or other radiographic imaging studies of the chest. These features can vary depending on the stage of infection in the lungs. In the early stages of PTB, the radiographic features may be minimal or not apparent at all. As the disease progresses, characteristic changes can appear on the X-ray, such as small nodules or infiltrates, representing an active lung infection. In more advanced stages of PTB, radiographic features such as cavitation (formation of cavities or holes in the lung tissue) and fibrosis (scarring of lung tissue) may be seen [1,6,11]. Therefore, the radiographic features of PTB can manifest in different ways depending on the stage of infection in the lungs. Radiologists and other healthcare professionals need to be familiar with these features to aid in the diagnosis and management of PTB.

To systematically examine chest X-rays, radiologists analyze various anatomical structures to identify any indications of disease or injury. This process includes conducting a technical quality check of the X-ray image, reviewing the patient's medical history, and examining the bones, lungs, heart and mediastinum, pleura, and diaphragm for any signs of abnormality. The radiologists then interpret their findings and prepare a report for the referring physician to aid in diagnosis and treatment planning. Similarly, a computer-aided system can be developed to detect radiographic features in chest X-ray images. Automatic computer-aided PTB detection from chest X-ray works by using deep learning algorithms to analyze the images and identify patterns indicative of PTB. Deep Learning can perform two tasks in this regard: classification and PTB manifestation localization. In the classification task, the ML models categorize CXR images into different classes of disease manifestation, whereas the localization tries to localize the pathology in the image. The process of localizing tuberculosis on chest X-rays involves recognizing and pinpointing the disease's pathology by examining the chest X-ray images.

The accurate diagnosis of PTB relies heavily on identifying its radiographic features in a chest X-ray, which can be achieved through semantic segmentation. Although fully supervised semantic segmentation with deep CNNs is becoming more prevalent, pixel-level annotation is required for this task, which is costly and challenging to obtain due to limited medical imaging data and domain expertise. To address this, a model that requires minimal pixel-by-pixel annotation data must be developed. Weakly supervised semantic segmentation offers a solution by enabling the acquisition of pixel-level labels (semantic segmentation) from image-level labels, thus overcoming these challenges. Despite the notion that employing weak localization techniques to detect tuberculosis from chest X-rays may appear as a new idea, a significant amount of research has been published on the subject.

The paper is organized as follows: Sect. 2 presents contributions and limitations of related works. Section 3 describes the proposed approach, the dataset preparation, evaluation methods, and experimental setup. Section 4, 5, and 6 present the findings, summarize the limitations of current weakly supervised techniques, and discuss future research directions respectively.

2   Related Work

In the field of weakly supervised segmentation, researchers have utilized the Semantic Texton Forest (STF) architecture and extended it for multiple-instance learning [33]. Additionally, some researchers have explored the use of an attention-based mechanism to understand better the contribution of each instance to the bag label [15]. These approaches aim to improve the accuracy and effectiveness of weakly supervised segmentation methods.

Qi et al. [23] proposed a deep learning model that is regularized with Graph Regularizer Embedding Network (GREN). They aimed to localize radiographic features on chest X-ray images by modeling the anatomical structure of the chest. GREN uses cross-image and cross-region information to capture relationships between and within images and computes graphs using regularizers such as Hash coding and Hamming distance to preserve structural information in the deep embedding layers of the neural network.

The most common approach for localizing the TB consistent region involves using visualization techniques to achieve weak localization, where radiographic features are localized by extracting visualizations from a classification model. Various visualization techniques, such as Occlusion Sensitivity [36], Saliency Maps [29], Class Activation Map [37], class-selective relevance maps (CRM)

[16], and Attention Networks [10], are commonly employed for the localization of pulmonary TB in Chest X-rays. These methods are also useful for interpreting complex models designed for medical fields [2, 4, 9].

Seeding-based weakly supervised segmentation is one of the prominent techniques and in some cases, it can even bring the best results. Classification neural networks can provide weak object localization but cannot accurately predict the spatial extent of objects. In this approach, the seed cues are generated by some visual explanations from deep networks via gradient-based localization or some other weak localization [17, 28, 32]. Another localization network can be built to benefit from the cues and provide a more accurate localization.

Kolesnikov et al. [17] trained a convolutional neural network (CNN) that can represent the likelihood of detecting labels at any position using a three-term loss function. This loss function consists of guidelines for localization, penalties for incorrect object predictions, and encouragement for segmentation that respects the spatial and color structure of images.

Tang et al. [32] suggested a principle called curriculum learning [3], which involves mastering knowledge in increasing order of difficulty. The model is initially trained on simpler examples and gradually progresses to more complex ones. In localizing the TB-consistent region, the approach uses CAM [37] to generate a heatmap, which is then refined using attention-guided iterative methods. By incorporating heatmap regression into the model, attention can guide the learning of superior convolutional features, resulting in heatmaps with increased significance.

The summary of the related works is presented in Table 1. These works collectively demonstrate the versatility of weakly supervised segmentation techniques, emphasizing the importance of attention mechanisms, regularizers, and curriculum learning for accurate and meaningful segmentation in medical imaging, particularly in localizing tuberculosis-related regions in chest X-rays. The major takeaways underscore the significance of attention-guided learning, leveraging regularizers for anatomical structure understanding, and employing seeding-based approaches and curriculum learning to enhance weakly supervised segmentation methodologies in medical image analysis. The motivation behind these efforts lies in improving the accuracy and interpretability of segmentation models in medical contexts, ultimately aiding in the effective diagnosis and treatment of diseases such as tuberculosis.

3   Methodology

3.1   Dataset

Chest X-rays are commonly used for identifying pulmonary tuberculosis, and various computer-aided detection methods have been developed using CXR images in recent years. This section will cover the detailed preparation and processing of the data.

The study used a publicly available chest X-ray dataset which is called TBX11K [19], this dataset is widely used in the research community and it

Table 1. Summary of Related Works.

Article/Year	Contribution	Limitation
Tang et al. [32], 2018	Take advantage of severity-level characteristics extracted from radiological reports	If the severity level identification is incorrect, the error will trickle down to the subsequent tasks and have an impact on overall performance
Sedai et al. [27], 2018	Utilized intermediate feature maps from the CNN layers at various deep network levels using class-aware deep-scale feature learning	Compared to the other similar studies, the obtained localization result is not significant
Liu et al. [18], 2019	Introduced modules for automatic image alignment in the direction of canonical alignment, and created a branch for contrast-induced attention that draws attention to every class of diseases from a pair of positive and negative images	The proposed network is complex and it can increase the resource requirements for training and inference time
Ouyang et al. [21], 2020	Presented a visual attention system with three levels (foreground, positive, and abnormality) that is hierarchical and allows for the gradual identification of the desired abnormality site	The resulting localization result is not significant when compared to other comparable investigations
Qi et al. [23], 2021	Take the cross-region and cross-image relationships into consideration using GREN	The lung lobes segmentation error may propagate and impact the performance of the localization

contains bounding box annotations, which are useful for weak localization. It has around 11, 200 images with 512×512 size. There are box annotations for 800 images which counts to 7.1% of the total sample. The dataset includes three classes: those with tuberculosis, those who are ill but not affected by the disease, and those in good health. The distribution of instances per class is illustrated in Table 1. The dataset is randomly divided into training, testing, and validation sets with ratios of 80%, 10%, and 10%, respectively.

Deep learning involves two important tasks: segmentation and classification. Classification refers to the problem of categorizing an entire entity into one of several predefined categories. Segmentation refers to the problem of dividing an image into parts and then classifying those parts. For example, consider an image of a chest X-ray. Segmentation would involve identifying tuberculosis pathology in the X-ray. Deep learning models are often used for segmentation and classification tasks because they can obtain hierarchical feature representations directly from the images without requiring handcrafted features.

Semantic segmentation is one of the deep learning techniques that can be used to recognize the radiographic features of Pulmonary TB in a chest

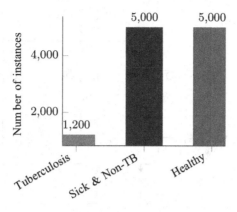

Fig. 1. Number of samples for each class.

X-ray. However, pixel-level annotation is required for this pixel-level classification task, which is expensive. Weakly supervised semantic segmentation helps to overcome these issues by providing a technique for obtaining pixel-level labels (semantic segmentation) from image-level labels. This technique is called weakly supervised because it uses image-level labels instead of pixel-level labels. This means that the algorithm does not need to know the exact location of each object in the image. Instead, it only needs to know which objects are present in the image. Many techniques have been applied and proven useful in this regard [18,21,22,27,30,32,35]. Most of these techniques use attention-based techniques, either in a hierarchical or multi-scale strategy.

Attention-based techniques are a type of deep learning algorithm that allows models to focus on specific parts of an input. This can be useful for tasks such as computer vision, where the model needs to understand the importance of a visual context. Hierarchical attention is a type of attention-based technique that allows models to focus on different levels of input. For example, a model could use hierarchical attention to focus on the individual parts of the image, as well as the relationships between those regions. Multi-scale attention is another type of attention-based technique that allows models to focus on different scales of an input. For example, a model could use multi-scale attention to focus on smaller and larger regions of the image.

The main purpose of this study is to localize a region that is affected by tuberculosis by using image level and a few pixel-level bounding box annotations. First, we developed a CNN for classification and we utilized the layers at different levels of the network and applied layer-wise learning and multi-scale attention. The model is then trained using image-level annotations and a few bounding box annotations. As depicted in Fig. 2, we introduce a framework for multi-scale feature learning for localization and classification. We chose a network architecture with 121 layers called DenseNet-121 [14]. It has 4 dense blocks with varying numbers of layers. The first dense block has 6 layers repeated twice, the second dense block has 12 layers repeated 12 times, the third dense block has

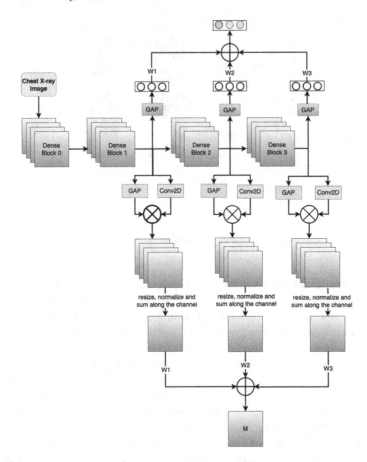

Fig. 2. Multi-scale attention for weak localization and classification of tuberculosis having DenseNet-121 [14] as a backbone network.

24 layers repeated 24 times and the fourth dense block has 16 layers repeated 16 times. Each layer in the dense block is connected to all the preceding layers by iterative concatenation of previous feature maps. This allows all layers to access feature maps from their preceding layers which encourages heavy feature reuse. The feature maps at the end of each block are down-sampled and passed to the next block. The global average pooling response of the feature maps from the final block is connected to a densely connected network to obtain the classification scores.

Classification-CNN. To classify efficiently, we used feature maps from the last three blocks of the DenseNet-121 model. We then obtained their Global Average Pooling (GAP) and fed each GAP to another Dense layer with neuron numbers equal to the class number. Finally, we used the Convex combination to calculate the weighted sum of the three dense layers and passed it through a sigmoid function. The Convex combination searches for the optimum weight to perform the weighted sum over the Dense layers. These weights are learned through back-propagation and are initialized with a weight equal to one divided by the number of classes ($\frac{1}{\|C\|}$).

Pathology Localization by Attention CNN (A-CNN). To perform pathology localization, we extended the classification network discussed in the previous section. We added another network (depicted in Fig. 3) to blocks 2, 3, and 4. The feature map from the selected levels is fed to the network built for localization purposes. The network takes in the feature maps to get its global average pooling (GAP) and its convolution with a kernel size of 1. We perform element-wise multiplication between the GAP value and the result from the convolution. The resulting product is then resized and normalized. Finally, we sum it along the channel axis to get single-channel images. We combine the results from each block using a convex combination and apply the sigmoid function. A convex combination is a linear combination of two or more values that produces another value within the same range. It is used in deep learning to combine the outputs of multiple models or classifiers.

Loss Functions. Two loss functions were applied to train the network: classification and localization loss. The classification loss is a categorical cross-entropy loss (Eq. 1) that measures the accuracy of the network's predictions for the manifestation classes. The localization loss (Eq. 2) is calculated from the bounding box and the class activation map obtained from the localization branch of the network. The two loss functions are combined using a weighted sum, and the weights of the two loss functions can be adjusted to trade-off between the accuracy of the classification and the localization. The use of two loss functions has been shown to improve the accuracy of object detection models, as it allows the network to learn both the classification and localization of PTB manifestations.

$$Loss_{cls} = \sum_k -y_k log(p_k) - \sum_k (1 - y_k)log(1 - p_k) \tag{1}$$

Where k is the class label, $k \in \{1, 2, ..., C\}$, C is the number of classes, y_k is the ground truth label, and p_k is the predicted label.

$$Loss_{loc} = \frac{1}{N} \sum_{y_n=1} \left(\frac{\sum_{ij}(T(M_{ij}^n) - G_{ij}^n)^2}{\sum_{ij} T(M_{ij}^n) + \sum_{ij} G_{ij}^n} \right) \tag{2}$$

where the image's ground-truth label $y_n \in \{0,1\}$ (0/1 indicates absence/presence of bounding box for the nth image), N is the number of images in the training batch, i and j represent the $(i,j)^{th}$ pixel in the corresponding M_{ij}^n and G_{ij}^n. G_{ij}^n is the mask created for the nth image through the bounding box, where the pixels inside the box are set to 1, while those outside are set to 0, and $T(M_{ij}^n)$ is the soft masking operation defined in Eq. 3 which masks out the impact of background noise in the attention maps.

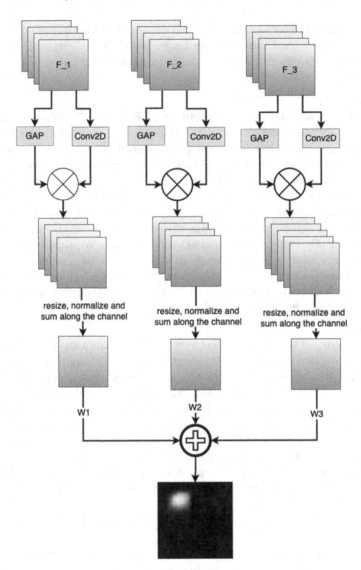

Fig. 3. Weak localization with multi-scale attention network.

$$T(M) = \frac{1}{1 + exp(-M)} \tag{3}$$

3.2   Evaluation Methods

In the development of a deep learning model for a given task, it is crucial to evaluate the model's performance to assess its effectiveness. In this study, we performed two main tasks: classification and localization. Therefore, we need to develop a method to measure the performance of the proposed model for both tasks.

For the classification task, we used four metrics to evaluate the performance of our model: accuracy, recall, precision, and F_1 score. Accuracy is the percentage of instances that are correctly classified, while recall is the percentage of positive instances that are correctly classified. Precision is the percentage of instances that are classified as positive that are actually positive. F_1 score is a harmonic means of precision and recall.

For the localization task, we used two metrics to evaluate the performance of our model: intersection over union (IoU) and Dice coefficient (DC). IoU (Eq. 4) is a measure of how well the predicted bounding box overlaps with the ground truth bounding box, while DC (Eq. 5) is a measure of how similar the predicted bounding box is to the ground truth bounding box.

$$IoU = \frac{\text{Overlap Area}}{\text{Predicted Mask} + \text{Ground Truth Mask} - \text{Overlap Area}} \tag{4}$$

$$DC = \frac{2 * \text{Overlap Area}}{\text{Predicted Mask} + \text{Ground Truth Mask}} \tag{5}$$

To assess the detection performance of the weak localization method, we employ intersection over union with various thresholds. For instance, using a threshold of 0.1 denoted by T(IoU) = 0.1 implies that instances with IoU values above 0.1 are classified as detected, while those equal to or below 0.1 are considered as missed detections. Consequently, the detection accuracy can be calculated by categorizing instances with IoU above 0.1 as detected and the rest as missed diagnoses Eq. 6. The same evaluation techniques exist in [18,27,32].

$$\text{Detection Accuracy} = \frac{\text{Count of instances with IoU exceeding 0.1}}{\text{Total instances with a bounding box}} \tag{6}$$

3.3 Experimental Setup

Using the ADAM optimizer, the model is trained on the dataset discussed in Sect. 3.1, TBX11K. The learning rate starts from 0.0001 and decreases by 10 times if the learning curve gets plateaued for two consecutive iterations. All the weights are initialized using a uniform. The mini-batch size is set to 4 with the NVIDIA TESLA T4 GPU. All the algorithms are implemented with Keras which runs on top of TensorFlow. The threshold of 0.5 is used to distinguish positive grids from negative grids in the class-wise feature map, which has been adopted in previous studies.

4   Result and Discussion

We conducted experiments to evaluate the performance of our model for both classification and localization. For classification, we used accuracy and F_1 score as our evaluation metrics. We used IoU for localization to measure the overlap between the predicted localization and the ground truth bounding box. We considered the pathology to be found if the IoU between the box and the predicted localization is greater than or equal to 0.1, $T(IoU) > 0.1$. However, it is important to note that the IoU threshold can be adjusted to reflect different levels of accuracy requirements.

The model achieved 98% accuracy and 97% F_1 score for classification and 83% accuracy at $T(IoU) = 0.1$ for localization. This means that the model can provide a prediction with an overlap ratio of 0.1 (which means that the predicted bounding box must overlap with the ground truth bounding box by at least 10% as depicted in Fig. 4 for 83% of the CXR images in the test set. Assessment of the proposed approach relative to other published techniques using classification and detection metrics is provided in Table 2.

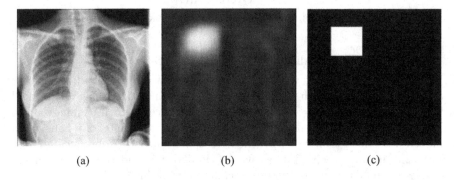

(a)   (b)   (c)

Fig. 4. a) original image b) generated localization c) the ground truth bounding box.

Table 2. Comparison of the proposed approach with other published techniques using classification and localization metrics.

Authors	Classification			Localization		
	Accuracy	AUC	F1 score	IoU	$T(IoU) = 0.1$	$T(IoU) = 0.5$
Rajaraman et al. [24]	92	96	94	62	–	–
Liu et al. [19]	90	–	–	–	–	–
Liu et al. [20]	93	98	–	31	–	74
Rajaraman et al. [25]	–	–	–	29	–	–
Proposed	**98**	**97**	**98**	**56**	**83**	–

Table 2 demonstrates that the classification performance surpasses the state-of-the-art study conducted on TBX11K. Additionally, the proposed method exhibits superior weak localization performance compared to previous studies utilizing weak localization techniques [20,25]. Notably, the localization performance is remarkably close to the fully supervised semantic segmentation proposed in [24], with only a 6% difference.

5   Conclusion

Deep learning models are effective for the localization of TB manifestation in CXR because they can automatically extract relevant features from the images without requiring manual feature engineering. By using multiple layers of non-linear transformations, deep learning models can capture complex and hierarchical patterns in the data and achieve high accuracy and generalization. However, training the fully supervised DL models to perform semantic segmentation requires a lot of pixel-level annotation data. Hence, an approach that utilizes a few bounding box annotations is required to alleviate the problem with fully supervised segmentation models. In some approaches, we can even get a promising result without using any annotation at all, by only considering the image-level labeling. However, having a few pixel-level annotations can significantly improve the performance of these approaches.

6   Future Work

In this study, we proposed an approach to localize TB manifestations in CXR images by using a DenseNet-121 model as a backbone. However, this model has a large number of parameters, which can lead to long inference times and high memory usage on chest X-ray devices. In future work, we plan to reduce the inference time and memory usage of the model by using a backbone network with fewer parameters. Additionally, we plan to expand our dataset to include other lung diseases, so that our approach can be used to diagnose a wider range of conditions.

References

1. Al Ubaidi, B.: The radiological diagnosis of pulmonary tuberculosis (tb) in primary care. RadioPaedia **4**, 73 (2018)
2. Ayano, Y.M., Schwenker, F., Dufera, B.D., Debelee, T.G.: Interpretable machine learning techniques in ECG-based heart disease classification: a systematic review. Diagnostics **13**(1), 111 (2022). https://doi.org/10.3390/diagnostics13010111
3. Bengio, Y., Louradour, J., Collobert, R., Weston, J.: Curriculum learning. In: Proceedings of the 26th Annual International Conference on Machine Learning, pp. 41–48 (2009)
4. Bonyani, M., Yeganli, F., Yeganli, S.F.: Fast and interpretable deep learning pipeline for breast cancer recognition. In: 2022 Medical Technologies Congress (TIPTEKNO), pp. 1–4 (2022)
5. Burrill, J., Williams, C.J., Bain, G., Conder, G., Hine, A.L., Misra, R.R.: Tuberculosis: a radiologic review. Radiographics **27**(5), 1255–1273 (2007)

6. Caws, M., Marais, B., Heemskerk, D., Farrar, J.: Tuberculosis in adults and children (2015)
7. Chakaya, J., et al.: Global tuberculosis report 2020-reflections on the global tb burden, treatment and prevention efforts. Int. J. Infect. Dis. **113**, S7–S12 (2021)
8. Chakaya, J., et al.: The who global tuberculosis 2021 report–not so good news and turning the tide back to end tb. Int. J. Infect. Dis. (2022)
9. Dasanayaka, S., Shantha, V., Silva, S., Meedeniya, D., Ambegoda, T.: Interpretable machine learning for brain tumour analysis using MRI and whole slide images. Softw. Impacts **13**, 100340 (2022). https://doi.org/10.1016/j.simpa.2022.100340
10. Ding, F., et al.: Hierarchical attention networks for medical image segmentation. arXiv preprint arXiv:1911.08777 (2019)
11. Niknejad, M., Gaillard, F.: Tuberculosis (pulmonary manifestations). J. Fam. Med. Dis. Prev. (2022). https://doi.org/10.53347/rID-8631
12. Hoog, A., et al.: A systematic review of the sensitivity and specificity of symptom and chest radiography screening for active pulmonary tuberculosis in hiv-negative persons and persons with unknown hiv status (2013). https://doi.org/10.13140/RG.2.2.19848.06406
13. van't Hoog, A.H., et al.: Screening strategies for tuberculosis prevalence surveys: the value of chest radiography and symptoms. PloS One **7**(7), e38691 (2012)
14. Huang, G., Liu, Z., Van Der Maaten, L., Weinberger, K.Q.: Densely connected convolutional networks. In: Proceedings of the IEEE Conference on Computer Vision and Pattern Recognition, pp. 4700–4708 (2017)
15. Ilse, M., Tomczak, J., Welling, M.: Attention-based deep multiple instance learning. In: International Conference on Machine Learning, pp. 2127–2136. PMLR (2018)
16. Kim, I., Rajaraman, S., Antani, S.: Visual interpretation of convolutional neural network predictions in classifying medical image modalities. Diagnostics **9**(2), 38 (2019)
17. Kolesnikov, A., Lampert, C.H.: Seed, expand and constrain: three principles for weakly-supervised image segmentation. In: Leibe, B., Matas, J., Sebe, N., Welling, M. (eds.) ECCV 2016. LNCS, vol. 9908, pp. 695–711. Springer, Cham (2016). https://doi.org/10.1007/978-3-319-46493-0_42
18. Liu, J., Zhao, G., Fei, Y., Zhang, M., Wang, Y., Yu, Y.: Align, attend and locate: chest x-ray diagnosis via contrast induced attention network with limited supervision. In: Proceedings of the IEEE/CVF International Conference on Computer Vision, pp. 10632–10641 (2019)
19. Liu, Y., Wu, Y.H., Ban, Y., Wang, H., Cheng, M.M.: Rethinking computer-aided tuberculosis diagnosis. In: Proceedings of the IEEE/CVF Conference on Computer Vision and Pattern Recognition, pp. 2646–2655 (2020)
20. Liu, Y., Wu, Y.H., Zhang, S.C., Liu, L., Wu, M., Cheng, M.M.: Revisiting computer-aided tuberculosis diagnosis. arXiv preprint arXiv:2307.02848 (2023)
21. Ouyang, X., et al.: Learning hierarchical attention for weakly-supervised chest x-ray abnormality localization and diagnosis. IEEE Trans. Med. Imaging **40**(10), 2698–2710 (2020)
22. Pan, C., et al.: Computer-aided tuberculosis diagnosis with attribute reasoning assistance. In: Wang, L., Dou, Q., Fletcher, P.T., Speidel, S., Li, S. (eds.) MICCAI 2022. LNCS, vol. 13431, pp. 623–633. Springer, Heidelberg (2022). https://doi.org/10.1007/978-3-031-16431-6_59
23. Qi, B., et al.: Gren: graph-regularized embedding network for weakly-supervised disease localization in x-ray images. arXiv preprint arXiv:2107.06442 (2021)

24. Rajaraman, S., Folio, L.R., Dimperio, J., Alderson, P.O., Antani, S.K.: Improved semantic segmentation of tuberculosis-consistent findings in chest x-rays using augmented training of modality-specific u-net models with weak localizations. Diagnostics **11**(4), 616 (2021)
25. Rajaraman, S., Guo, P., Xue, Z., Antani, S.K.: A deep modality-specific ensemble for improving pneumonia detection in chest x-rays. Diagnostics **12**(6), 1442 (2022)
26. Ryu, Y.J.: Diagnosis of pulmonary tuberculosis: recent advances and diagnostic algorithms. Tubercul. Respirat. Dis. **78**(2), 64–71 (2015)
27. Sedai, S., Mahapatra, D., Ge, Z., Chakravorty, R., Garnavi, R.: Deep multiscale convolutional feature learning for weakly supervised localization of chest pathologies in x-ray images. In: International Workshop on Machine Learning in Medical Imaging, pp. 267–275 (2018)
28. Selvaraju, R.R., Cogswell, M., Das, A., Vedantam, R., Parikh, D., Batra, D.: Gradcam: visual explanations from deep networks via gradient-based localization. In: Proceedings of the IEEE International Conference on Computer Vision, pp. 618–626 (2017)
29. Simonyan, K., Vedaldi, A., Zisserman, A.: Deep inside convolutional networks: visualising image classification models and saliency maps. arXiv preprint arXiv:1312.6034 (2013)
30. Singh, A., et al.: Deep learning for automated screening of tuberculosis from Indian chest x-rays: analysis and update. arXiv preprint arXiv:2011.09778 (2020)
31. Steingart, K.R., et al.: Xpert® mtb/rif assay for pulmonary tuberculosis and rifampicin resistance in adults. Cochrane Database System. Rev. (2013)
32. Tang, Y., Wang, X., Harrison, A.P., Lu, L., Xiao, J., Summers, R.M.: Attention-guided curriculum learning for weakly supervised classification and localization of thoracic diseases on chest radiographs. In: Shi, Y., Suk, H.-I., Liu, M. (eds.) MLMI 2018. LNCS, vol. 11046, pp. 249–258. Springer, Cham (2018). https://doi.org/10.1007/978-3-030-00919-9_29
33. Vezhnevets, A., Buhmann, J.M.: Towards weakly supervised semantic segmentation by means of multiple instance and multitask learning. In: 2010 IEEE Computer Society Conference on Computer Vision and Pattern Recognition, pp. 3249–3256. IEEE (2010)
34. Vonasek, B., et al.: Screening tests for active pulmonary tuberculosis in children. Cochrane Database System. Rev. (2021)
35. Wang, X., Peng, Y., Lu, L., Lu, Z., Summers, R.M.: Tienet: text-image embedding network for common thorax disease classification and reporting in chest x-rays. In: Proceedings of the IEEE Conference on Computer Vision and Pattern Recognition, pp. 9049–9058 (2018)
36. Zeiler, M.D., Fergus, R.: Visualizing and understanding convolutional networks. In: European Conference on Computer Vision, pp. 818–833 (2014)
37. Zhou, B., Khosla, A., Lapedriza, A., Oliva, A., Torralba, A.: Learning deep features for discriminative localization. In: Proceedings of the IEEE Conference on Computer Vision and Pattern Recognition, pp. 2921–2929 (2016)

Automated Kidney Segmentation and Disease Classification Using CNN-Based Models

Akalu Abraham[1]([✉])[iD], Misganu Tuse[1][iD], and Million Meshesha[2][iD]

[1] Addis Ababa Science and Technology University, Addis Ababa, Ethiopia
ake.abrish@gmail.com
[2] School of Information Science, Addis Ababa University, Addis Ababa, Ethiopia

Abstract. According to studies, kidney disease (KD) is predicted to be the 5th leading cause of death by 2040. Some works have been done on KD segmentation and classification using CNN-based models. However, works that consider automated KD segmentation and classification using CNN-based models for more than one KD type, together with model optimization are rarely investigated. In this study, four KD conditions (tumor, cyst, stone, and normal) are considered to develop automated kidney segmentation and disease classification using CNN-based models. We utilized a balanced and limited number of images from a publicly available CT scan image dataset containing 12,446 images categorized into four classes (tumor, cyst, stone, and normal). The overall model development procedure has four major stages. The first stage, the data preprocessing stage, involves target mask preparation and data augmentation. The second stage involves training a U-Net segmentation model using the augmented dataset. In the third stage, CNN-based models, VGG-16, ResNet-50, and Inception-V3, were trained on segmented images generated using a segmentation model to classify KD. In this stage, batch normalization, dropout, regularization, and dense layers are added to optimize the models. Finally, the model is designed to provide possible expert-level treatment recommendations. The experiment results confirmed that using the threshold technique improved the performance of the U-Net model with an accuracy of 98%, and a Dice coefficient of 0.9. All the models were trained with the same amount of dataset and fixed hyperparameters; however, ResNet-50 outperformed with a testing accuracy of 97%.

Keywords: UNet segmentation · Image classification · Automated disease diagnosis · CT scan images

1 Introduction

The kidneys play a vital role in filtering waste products from the blood [1]. Additionally, the kidneys maintain fluid and mineral balance, produce vitamin D, regulate blood pressure, and stimulate red blood cell production, collectively

T. G. Debelee et al. (Eds.): PanAfriConAI 2023, CCIS 2068, pp. 60–72, 2024.
https://doi.org/10.1007/978-3-031-57624-9_3

contributing to a person's overall health [2]. Kidney disease, however, disrupts these functions, delaying the body's ability to cleanse the blood, remove excess water, and maintain blood pressure [3]. According to studies in 2019, KD was the 10th leading cause of death, and it is predicted to rise to the 5th position by 2040 [4,5].

Common types of kidney diseases include cyst formation, nephrolithiasis (kidney stones), and renal cell carcinoma (kidney tumor), collectively impacting kidney function and health [6]. Notably, kidney disease often progresses silently in its early stages, with no apparent symptoms, rendering it potentially fatal. In many cases, individuals may lose up to 90% of their kidney function before manifesting any symptoms, leaving them unaware of their condition [7,8]. Nevertheless, early detection and timely treatment can significantly enhance the chances of prevention and cure.

To diagnose kidney disease, medical professionals rely on a combination of medical images, blood tests, and urine analyses. Among various imaging modalities, computed tomography (CT) scans are particularly ideal for studying kidney diseases due to their ability to provide three-dimensional information and detailed, slice-by-slice images. These CT images are valuable resources for implementing machine learning models for automated kidney disease analysis [3,9]. Currently, the availability of powerful computing resources and extensive medical datasets has revolutionized deep learning techniques. Deep learning plays a pivotal role in minimizing human errors, ultimately enhancing the performance of the models, and facilitating early disease detection [10].

Among deep learning models, Convolutional Neural Networks (CNNs), CNN-based pre-trained models, and Vision Transformers (ViTs) have been extensively applied in kidney segmentation and disease classification. These models have demonstrated superior accuracy and diagnosis efficiency [9]. Identifying and categorizing KD based on CT scan images without segmentation poses a tough challenge, given that CT scans produce intricate three-dimensional representations encompassing not only the kidneys but also the adjacent tissues and blood vessels. The complexities of these images demand accurate segmentation for proper disease classification and diagnosis.

There have been works done on kidney disease segmentation and classification using CNN-based models [11–13]. However, these studies have predominantly focused on identifying single disease types, most commonly kidney tumor [1,2,11]. Works that consider automated kidney disease segmentation and classification using CNN-based models for all types of kidney diseases, along with model optimization, are rarely investigated. Thus, this research aimed to develop kidney CT scan image segmentation and disease classification using pre-trained CNN-based models for the most common types of kidney diseases. By enabling fast, accurate diagnosis of kidney diseases and reducing the cost of diagnosis, the work promises to enhance healthcare delivery, provide valid decision-making tools for medical professionals, and mitigate the scarcity of specialists.

2   Related Works

Various authors showed that deep learning has the potential to revolutionize medical image analysis and diagnosis, as well as improve patient care and outcomes [3,8,14,15]. Recently, Mehedi et al. [12] presented a segmentation-based kidney tumor classification method using CNN-based Deep Neural Network (DNN). The authors proposed deep learning-based segmentation models (U-Net along with SegNet) after segmenting kidneys using manual segmentation techniques. Then, for the classification task, they trained the modified MobileNetV2, VGG16, and InceptionV3 on the segmented kidney data.

Similarly, Pan et al. [13] presented segmentation and classification of the renal tumor, intended for preoperative assessment of renal tumor. Further, they implemented a 2-step segmentation strategy to the segmentation module and improved the result by 2.8%. The experimental results of classification and segmentation reported 100% accuracy and 0.882 dice coefficient of tumor region respectively and concluded that the results are better than a single classification network and segmentation network.

A study by Z. Gonga and L. Kanb [11] presented a model based on CNN, that can segment the ROI and classify kidney tumor at the same time. They proposed a multi-task network architecture for segmentation and classification, 2D SCNet. In this study, authors designed the multi-task network by providing/feeding global features in the classification stage, while the segmentation stage focused on the local features and ROI. They incorporated ResNet-50 as a base network for feature sharing, which provided better performance compared to Dense Net. Their multi-stage model achieved a classification accuracy of 99.5% in both classes (benign and malignant) whereas the segmentation network reached a dice coefficient of 0.946 and 0.846 respectively. They also compared with other network architectures, PSPNet and their proposed network, 2D SCNet achieved 4.9% improvement.

In [2] authors proposed a kidneys and renal mass diagnosis framework incorporating 3D segmentation and renal mass subtype classification using the basic 3D U-Net structure, with residual blocks included for gaining the cross-layer connections. Their framework is composed of two stages, the 3D segmentation stage (fully supervised manner) and the renal mass classification network, a CNN-based network (3D U-Net) that classifies into subtype classification. The model scored a Dice coefficient of 0.96 and 0.86 for kidney and renal mass regions, respectively.

Authors in [1] develop a model to improve the segmentation and classification of kidney tumor with small-size 3D CT images via transfer learning based on CNNs. In this approach, authors proposed a two-channel enhanced segmentation network architecture, FS-Net. They designed a network with two-channel data input that can segment kidney and tumor regions. After segmentation, they designed discriminative classifiers to further classify the tumor class according to the segmented regions. They used 3D-ResNet for feature extraction and SVM for tumor classification. They achieved a dice coefficient of 0.9662 for kidney segmentation and 0.7685 for tumor region segmentation.

3 Materials and Methods

3.1 Dataset Description

The dataset was collected from PACS (Picture archiving and communication system) at various hospitals in Dhaka, Bangladesh, where patients had already been diagnosed with kidney tumor, cysts, normal conditions, or kidney stones. This dataset is publicly available and was downloaded from kaggle.com. It comprises 12,446 unique images classified by domain experts. Accordingly, 3,709 images in the cyst class, 5,077 in the normal class, 1,377 in the stone class, and 2,283 in the tumor class. However, we were unable to use all of the data since we employed an automatic segmentation technique that required manual preparation of mask, making the process time-consuming.

3.2 Architecture of the Proposed Model

The model's architecture consists of four major phases: the Data preprocessing phase, the Segmentation phase; the Classification phase; and the Treatment Plan phase. See Fig. 1 below.

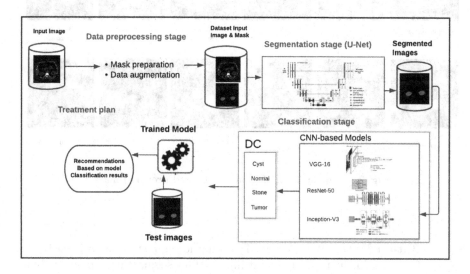

Fig. 1. Architecture of the proposed model

Data Preprocessing Phase: This phase involves target mask preparation and data augmentation.

- **Target Mask Preparation**: We prepared target masks for the input images using ITK Snap and ImageJ (Fiji). For this research, we manually created 100 target masks for each class, resulting in a total of 400 target masks for the four classes. Due to the tedious and time-consuming nature of manual work, we limited the number of target masks prepared and employed an augmentation technique to expand the dataset.

- **Data Augmentation**: In this stage, we augmented both the input images and target masks using the albumentations library. We utilized five different data augmentation techniques: CenterCrop, RandomRotate90, GridDistortion, HorizontalFlip, and VerticalFlip. This augmentation process increased the dataset to 2400 images and 2400 corresponding masks (400 original images and 400 masks, each augmented with the five techniques). For each input image and target mask, we generated five additional images and target masks as shown in the Fig. 2 below. Prior to training the segmentation model, we applied a threshold value of 0.5 to the augmented target masks to enhance it by reducing the impact of noise in the target mask.

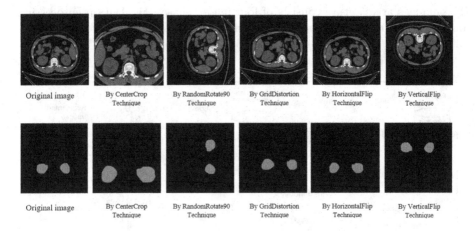

Fig. 2. Augmentation of input image and prepared mask

Segmentation Phase: We employed automatic segmentation to identify the region of interest in the provided input images. Our segmentation model is built upon the pre-trained U-Net model, developed by the University of Freiburg, department of computer science. It was developed for biomedical image segmentation. U-Net is a U-shaped CNN-based network containing a contractive path (encoder), bottleneck, expansive path (decoder), and skip-connection [16]. Our segmentation model takes two types of images as inputs: the input images themselves and manually prepared masks. The model's output is the predicted mask, which serves as the input for the subsequent classification model. The diagram below illustrates the three images: the input image, the manually prepared mask, and the predicted mask generated by our segmentation model (Fig. 3).

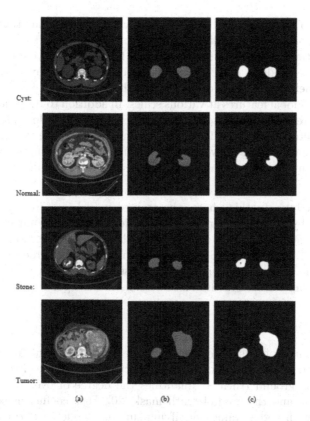

Fig. 3. Segmentation: (a) input images, (b) prepared mask, and (c) predicted mask.

Classification Phase: The classification model utilizes the predicted mask generated by the image segmentation model. To achieve classification, we employed various pre-trained models, including VGG-16, ResNet-50, and Inception-V3 to classify kidney CT scan images into four distinct categories: cyst, normal, stone, and tumor.

The VGG-16 model comprises 16 layers, encompassing 13 convolutional layers and 3 fully connected layers. The convolutional layers are organized into blocks, with each block containing multiple convolutional layers followed by a max-pooling layer. This network accepts input images of size 224×224 and has been trained on an extensive image dataset from the ImageNet database, which includes 1000 different output classes [17].

ResNet-50 (Residual Network 50) is a deep convolutional neural network architecture renowned for its depth. It comprises 50 layers, encompassing 49 convolutional layers and a single fully connected layer. What sets ResNet50 apart is its utilization of residual blocks, designed to combat the issue of vanishing gradients that can afflict extremely deep networks. A residual block is formed by integrating two or more convolutional layers, with the output of each layer being added to the input of the block. This creates a shortcut that facilitates easier backpropagation of gradients throughout the network [18].

The Inception-v3 architecture is renowned for its innovative use of Inception modules. These modules consist of multiple convolutional layers with diverse sizes and kernel shapes. This design allows the network to effectively capture both local and global features at various scales. In addition to Inception modules, the architecture incorporates several other techniques, including batch normalization and factorized convolutions. These techniques enhance the efficiency and overall performance of the network. In total, the Inception-v3 network comprises 42 layers, encompassing 39 convolutional layers and 3 fully connected layers [19].

Treatment Plan: In this phase, the trained model is stored in a file, prepared for use by domain experts, including doctors, clinical nurses, and other healthcare professionals. These experts can then import the model and input new kidney CT scan images. Subsequently, the model will proceed to segment and classify the images into their respective disease types, offering corresponding recommendations for each identified kidney disease category.

3.3   Evaluation Metrics

For the segmentation task, we employed the Dice coefficient as the primary evaluation metrics, which indicates the perfect overlap between the segmented region and the ground truth. It quantifies the degree of overlap between the predicted mask and the ground truth mask [20]. Dice coefficient scores range from 0 to 1, with higher values signifying superior model performance [21]. In the context of classification models, we adopted a train-test split approach to evaluate model performance. That is, dividing the dataset into three subsets: namely, training set, validation set, and testing set. We assessed the models using a set of key metrics such as accuracy, precision, recall, and f1-score.

4   Result and Discussion

The proposed model demonstrates high segmentation performance with an accuracy of 98% (0.9 Dice coefficient) and high disease classification performance with 97% accuracy on the test data. The result of the experiments indicates the effectiveness of this method in kidney segmentation and disease classification on CT images. The following quantitative results were obtained and used to evaluate the models' performance both in the segmentation and classification stages.

4.1   Segmentation Result

As explained in the data preprocessing phase under Sect. 3.2, we prepared 400 masks from 400 input images manually. These input images and masks were both augmented and increased the dataset to 2400 input images and 2400 corresponding masks (with a threshold value of 0.5) by using argumentation techniques from albumentations library. From this total dataset (2400 input images and 2400 corresponding masks), 80% (1920 input images and 1920 corresponding masks) were used for training, and 20% (480 input images and 480 associated masks) for validation and testing. Then, these data (input images and masks) were trained by using U-Net, a CNN-based pre-trained model. The images were resized to (256, 256, 3), which is the input size of U-Net. The Segmentation model was trained with 20 epochs and optimized by the Adam optimizer with a learning rate of 0.0001. We applied the binary cross-entropy for the loss function. The batch size was set to 16. We have used early stopping to prevent over-fitting of the model to the training data and to improve the generalization performance of the model. Finally, the output will be the segmented CT scan image, which is the predicted mask. The result of the experiment showed that the segmentation model achieved an accuracy of 98% and a Dice coefficient of 0.9.

4.2   Classification Result

The dataset (2640 masks generated from the segmentation stage) was split into 80% (2112 masks, 528 from each class) training set and 20% test set (528 masks, 132 from each class). Further, the training set was split into 20% validation set (528 masks, 132 from every class) and the rest number of the data (1,584 images, 396 from each class) were taken as a training set. Then all three models were trained with the same hyper-parameters. That is, the number of epochs was 50, Adam optimizer with a learning rate of 0.0001, batch size of 16, and categorical cross-entropy for computing the loss. Moreover, BatchNormalization, regularization, dropout, and dense layers are employed in all three models, which improved the accuracy of the models.

In order to evaluate/ measure the efficiency of a classification model accuracy, F1-score, recall, precision, accuracy, and confusion matrix are utilized, which are explained in Table 1 and Fig. 4, 5 and 6. Moreover, training and validation accuracy and corresponding loss curves are also presented in Fig. 7, 8 and 9 for all three models trained on a similar dataset. In Fig. 4, 5 and 6 the confusion matrices for all three cases are presented and it can be observed that ResNet-50 achieved perfect scores on both cyst and tumor classes but the accuracy dropped corresponding to the normal and stone classes; misclassified 14 out of 132 and 11 out of 132 respectively.

Table 1. Performance evaluation of the three CNN-based models.

Model Performance Evaluation				
Models	Accuracy	Precision	Recall	F1-score
VGG-16	92%	92%	92%	92%
ResNet-50	**97%**	**97%**	**97%**	**97%**
Inception-V3	91%	91%	91%	91%

In Fig. 7, 8 and 9 training and validation accuracy and corresponding loss curve for all three models are displayed for 50 epochs.

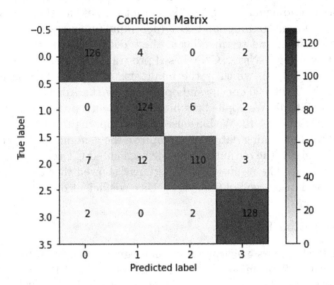

Fig. 4. Confusion matrix for VGG-16 based model

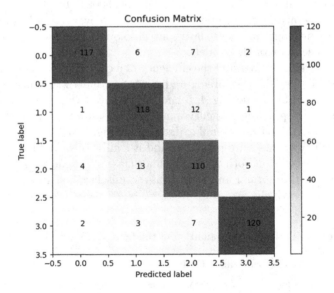

Fig. 5. Confusion matrix for InceptionV3 based model

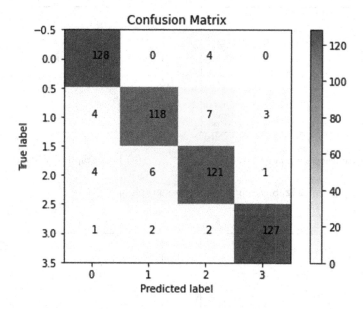

Fig. 6. Confusion matrix for ResNet-50 based model

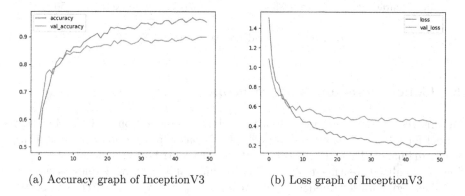

(a) Accuracy graph of InceptionV3   (b) Loss graph of InceptionV3

Fig. 7. Accuracy and Loss graph of InceptionV3

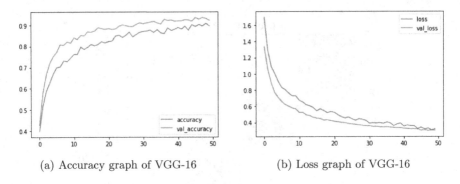

(a) Accuracy graph of VGG-16   (b) Loss graph of VGG-16

Fig. 8. Accuracy and Loss graph of VGG-16

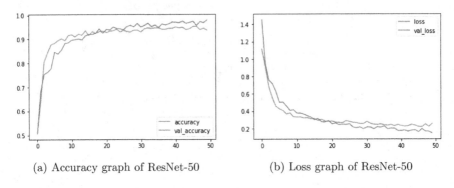

(a) Accuracy graph of ResNet-50   (b) Loss graph of ResNet-50

Fig. 9. Accuracy and Loss graph of ResNet-50

5   Conclusions and Future Scope

In this research, we have developed a robust segmentation and classification model for kidney diseases by leveraging CNN-based pre-trained networks using CT scan images. This model has the potential to contribute significantly to devising personalized treatment plans for individuals afflicted with kidney diseases.

The segmentation model, trained over 20 epochs with a batch size of 16, achieved an accuracy rate of 98%, a precision of 98.7%, and a Dice coefficient of 0.9. Notably, our experiments indicated that implementing a threshold technique significantly enhanced the performance of the U-Net model by reducing the impact of noise.

Subsequently, we utilized the output images generated by the trained U-Net model, which served as predicted masks to train three distinct CNN-based classification models: VGG-16, ResNet-50, and InceptionV3. All three models were trained using the same comprehensive dataset, maintaining consistent training, validation, and testing sets, along with identical hyperparameters. Nevertheless, each model exhibited varying degrees of performance. Notably, ResNet-50 emerged as the top performer, achieving an impressive testing accuracy of 97%. However, some misclassifications were observed in the testing data, with variations seen among the different disease classes. This is because of the dataset which contains certain CT scan images with multiple types of kidney diseases, but they are considered/labeled as one of the classes during training.

In summary, this research effectively demonstrates the utilization of deep learning techniques for the task of kidney segmentation and disease classification. Although the results are promising, further research and investigations are advised to enhance the model's performance and to address the challenges associated with single-image data containing more than one disease class at a time. Moreover, we recommend exploring multi-label disease classification, as single image may encompass multiple disease classes simultaneously.

References

1. Zhu, X.-L., Shen, H.-B., Sun, H., Duan, L.-X., Xu, Y.-Y.: Improving segmentation and classification of renal tumors in small sample 3d ct images using transfer learning with convolutional neural networks. Int. J. Comput. Assist. Radiol. Surg. **17**(7), 1303–1311 (2022)
2. Liu, J., Yildirim, O., Akin, O., Tian, Y.: Ai-driven robust kidney and renal mass segmentation and classification on 3d ct images. Bioengineering **10**(1), 116 (2023)
3. Gharaibeh, M., et al.: Radiology imaging scans for early diagnosis of kidney tumors: a review of data analytics-based machine learning and deep learning approaches. Big Data Cogn. Comput. **6**(1), 29 (2022)
4. Abdel-Fattah, M.A., Othman, N.A., Goher, N.: Predicting chronic kidney disease using hybrid machine learning based on apache spark. Comput. Intell. Neurosci. **2022** (2022)
5. Report, W.: The top 10 causes of death. WHO report, vol. 2020 (2020). https://www.who.int/news-room/fact-sheets/detail/the-top-10-causes-of-death
6. Hsiao, C.-H., et al.: A deep learning-based precision volume calculation approach for kidney and tumor segmentation on computed tomography images. Comput. Methods Programs Biomed. **221**, 106861 (2022)
7. Pal, S.: Chronic kidney disease prediction using machine learning techniques. Biomed. Mater. Dev. **1**(1), 534–540 (2023)
8. Dai, D., Alvarez, P.J., Woods, S.D.: A predictive model for progression of chronic kidney disease to kidney failure using a large administrative claims database. Clinicoecon. Outcomes Res. CEOR **13**, 475 (2021)
9. Islam, M.N., et al.: Vision transformer and explainable transfer learning models for auto detection of kidney cyst, stone and tumor from ct-radiography. Sci. Rep. **12**(1), 1–14 (2022)
10. Dritsas, E., Trigka, M.: Machine learning techniques for chronic kidney disease risk prediction. Big Data Cogn. Comput. **6**(3), 98 (2022)
11. Gong, Z., Kan, L.: Segmentation and classification of renal tumors based on convolutional neural network. J. Radiat. Res. Appl. Sci. **14**(1), 412–422 (2021)
12. Mehedi,M.H.K., Haque, E., Radin, S.Y., Ur Rahman, M.A., Reza, M.T., Alam, M.G.R.: Kidney tumor segmentation and classification using deep neural network on ct images. In: 2022 International Conference on Digital Image Computing: Techniques and Applications (DICTA), pp. 1–7 (2022)

13. Pan, T., et al.: A multi-task convolutional neural network for renal tumor segmentation and classification using multi-phasic ct images. In: 2019 IEEE International Conference on Image Processing (ICIP), pp. 809–813 (2019)
14. Bai, Q., Su, C., Tang, W., Li, Y.: Machine learning to predict end stage kidney disease in chronic kidney disease. Sci. Rep. **12**(1), 1–8 (2022)
15. Senan, E.M., et al.: Diagnosis of chronic kidney disease using effective classification algorithms and recursive feature elimination techniques. J. Healthcare Eng. **2021** (2021)
16. Ronneberger, O., Fischer, P., Brox, T.: U-net: convolutional networks for biomedical image segmentation. In: Navab, N., Hornegger, J., Wells, W.M., Frangi, A.F. (eds.) MICCAI 2015. LNCS, vol. 9351, pp. 234–241. Springer, Cham (2015). https://doi.org/10.1007/978-3-319-24574-4_28
17. Simonyan, K., Zisserman, A.: Very deep convolutional networks for large-scale image recognition. arXiv:1409.1556 (2014)
18. He, K., Zhang, X., Ren, S., Sun, J.: Deep residual learning for image recognition. In: Proceedings of the IEEE Conference on Computer Vision and Pattern Recognition, pp. 770–778 (2016)
19. Szegedy, C., Vanhoucke, V., Ioffe, S., Shlens, J., Wojna, Z.: Rethinking the inception architecture for computer vision. In: Proceedings of the IEEE Conference on Computer Vision and Pattern Recognition, pp. 2818–2826 (2016)
20. Long, J., Shelhamer, E., Darrell, T.: Fully convolutional networks for semantic segmentation. In: Proceedings of the IEEE Conference on Computer Vision and Pattern Recognition, pp. 3431–3440 (2015)
21. Taha, A.A., Hanbury, A.: Metrics for evaluating 3d medical image segmentation: analysis, selection, and tool. BMC Med. Imaging **15**(1), 1–28 (2015)

Generating Synthetic Brain Tumor Data Using StyleGAN3 for Lower Class Enhancement

Ahmed Abdalaziz[1,2] and Friedhelm Schwenker[1(✉)]

[1] University of Ulm Institute of Neural Information Processing, 89081 Ulm, Germany
friedhelm.schwenker@uni-ulm.de
[2] German University in Cairo, New Cairo 11835, Egypt
ahmed.ashrafmansour@student.guc.edu.eg

Abstract. Accurate simulation of brain tumor data is crucial in medical imaging research for developing and assessing novel image enhancement techniques. We propose an approach to generate synthetic brain tumor data using StyleGAN3, a GAN model. Our objective is to explore the impact of varying gamma values and dataset sizes on the quality of synthetic brain tumor images via transfer learning from a pre-trained 512 X 512 model.

We conducted three experiments with different gamma values and dataset sizes. The first two experiments employed gamma values of 6.6 and 8.2, respectively, using a dataset of 310 brain tumor images. The third experiment maintained the gamma value at 8.2 but reduced the dataset size to 155 images.

We used FID and KID as evaluation metrics. Results show varying gamma values and dataset sizes significantly influence synthetic image quality. Higher gamma values generally yielded improved image quality. In another experiment, we obtained FID and KID scores of 67.856 and 0.013, respectively [1].

These findings impact brain tumor imaging enhancement techniques. StyleGAN3 and transfer learning can efficiently generate synthetic brain tumor datasets, advancing medical image analysis and diagnosis. Our study underscores synthetic data generation's potential for enhancing medical image research.

Keywords: Brain Tumor Imaging · Evaluation Metrics · Image Enhancement · StyleGAN3 · Synthetic Data Generation · Transfer Learning

Acronyms

GAN	Generative Adversarial Network
AFHQ	Animal FacesHQ
FID	Frechet Inception Distance
KID	Kernel Inception Distance
MRI	Magnetic Resonance Imaging

T. G. Debelee et al. (Eds.): PanAfriConAI 2023, CCIS 2068, pp. 73–87, 2024.
https://doi.org/10.1007/978-3-031-57624-9_4

SSIM	Structural Similarity Index
DSC	Dice Similarity Coefficient
PSNR	Peak Signal-to-Noise Ratio
DCGAN	Deep Convolutional Generative Adversarial Network
AGGrGAN	Aggregation of GANs
CNN	Convolutional Neural Network
WGAN	Wasserstein GAN
AdaIN	Adaptive Instance Normalization

1   Introduction

Medical imaging is crucial for diagnosing and treating diseases like brain tumors. Accurate algorithms for tumor detection rely on diverse, high-quality data, but obtaining such data can be challenging due to privacy concerns, data scarcity, and class imbalance.

Class imbalance, with more common than rare tumor types, can bias machine learning models. Synthetic data generation, particularly using GANs, has shown promise in overcoming data limitations in various fields.

In medical imaging, generating synthetic brain tumor data, especially for less common types, is essential. This paper is motivated by the need for realistic and diverse synthetic brain tumor data for image enhancement techniques. It aims to investigate the impact of gamma values and dataset sizes on synthetic image quality using transfer learning from a pre-trained model.

The class imbalance problem is addressed by generating images representing both common and rare tumor types. This ensures model fairness and accuracy.

1.1   Challenges

In the pursuit of generating synthetic images for medical research, several challenges demand careful consideration and strategic solutions. One crucial obstacle involves the necessity to ensure both realism and diversity in the synthetic images produced. Striking the right balance between authenticity and variety is paramount to the success of the endeavor.

Another pressing issue that requires meticulous attention is the inherent class imbalance observed in brain tumor data. Tackling this challenge involves developing effective strategies to address and rectify the uneven distribution of data across different classes. This imbalance poses a significant hurdle that must be overcome to enhance the reliability and robustness of the generated synthetic images.

Additionally, a noteworthy goal in this domain is to achieve diversity not only in the overall dataset but specifically in terms of tumor types and stages. Successfully incorporating a wide range of tumor variations is essential for the comprehensive utility and applicability of synthetic images in medical research. This challenge necessitates a thoughtful and inclusive approach to ensure that the synthetic dataset captures the complexity of real-world scenarios, spanning diverse tumor types and their various developmental stages.

1.2 Objective

The primary objective of this research is to develop a robust methodology for generating synthetic brain tumor data using the StyleGAN3 architecture. Our key research goals encompass various aspects:

1. Adapting StyleGAN3 to the domain of medical imaging, specifically for brain tumor data generation.
2. Exploring transfer learning techniques to enhance the generative capability of StyleGAN3 using pre-trained models.
3. Conducting systematic experiments to evaluate the impact of different parameters on the quality and diversity of synthetic brain tumor images.
4. Employing quantitative evaluation metrics to assess the realism and clinical relevance of the generated data.

Through this paper, we aim to contribute to the advancement of synthetic data generation techniques for medical applications, ultimately aiding researchers, clinicians, and medical practitioners in developing and evaluating innovative solutions for brain tumor diagnosis and treatment.

This paper aims to develop a method for generating realistic, diverse and class-balanced synthetic brain tumor data for enhancing image analysis. The findings will have implications in medical imaging, advancing image enhancement techniques for brain tumor data.

2 Background and Literature Review on Generating Synthetic Brain Tumor Data Using GANs

Brain tumors are challenging to detect and diagnose, with MRI being a common but resource-intensive imaging technique. GANs have proven effective in generating synthetic data for various domains, including medical imaging.

In [2], used GANs to generate synthetic brain tumor images for training a classification model, achieving 98% accuracy. This is a significant improvement over previous methods, which typically achieve around 90% accuracy.

In [3], used GANs to create synthetic brain tumor images representing various tumor types and stages, resulting in a high Dice coefficient of 0.92 for tumor segmentation.

The use of GANs to generate synthetic medical images is a promising new approach that has the potential to improve the accuracy of medical diagnosis. However, there are some limitations to the existing literature, such as a limited focus on low-grade brain tumors, the use of limited datasets, and the lack of a standard method for evaluating the quality of synthetic images.

The present study addresses the limitations of existing research by employing StyleGAN3, the current state-of-the-art GAN architecture, for generating synthetic brain tumor images. To comprehensively evaluate the effectiveness of StyleGAN3, three distinct experiments were conducted. By incorporating Style-GAN3 and conducting a thorough evaluation, this study provides a valuable contribution to the field of synthetic brain tumor image generation. The findings demonstrate the potential of StyleGAN3 in producing high-quality, realistic synthetic brain tumor images that can effectively support medical diagnosis and treatment planning.

2.1 Medical Image Synthesis

Recent advancements in deep learning are limited by the availability of high-quality medical datasets due to privacy and data scarcity issues. Synthetic images offer a solution for medical image analysis, including brain tumor diagnosis.

Deep learning models require vast amounts of high-quality data for effective training. However, acquiring and labeling such datasets can be expensive, time-consuming, and challenging due to privacy concerns and the limited availability of labeled data. Moreover, datasets may inherently contain biases, leading to biased models that make inaccurate or unfair predictions [4].

Training deep learning models can be computationally demanding, requiring specialized hardware and software, such as high-performance GPUs and specialized libraries like TensorFlow or PyTorch. This computational overhead can limit the accessibility of deep learning to researchers and practitioners with limited resources [5].

2.2 Generative Adversarial Networks

GANs consist of a generator and discriminator network engaged in an adversarial training process. They have been instrumental in medical image synthesis and image-to-image translation, demonstrating an ability to generate lifelike images.

2.3 DCGAN

The DCGAN, in [8], introduces a novel approach employing a DCGAN architecture to generate realistic synthetic MRI scans of brain tumors. Leveraging the Brain Tumor Segmentation (BRATS) challenge dataset with over 2,000 MRI scans across glioblastoma (GBM), meningioma, and pituitary tumors, the DCGAN-MRI model encompasses a generator and a discriminator. The generator produces synthetic MRI scans from random noise vectors, compelling improvements through adversarial training against the discriminator [9].

Performance evaluation, utilizing metrics such as SSIM, PSNR, and DSC, showcased the DCGAN-MRI model's ability to generate highly realistic images closely resembling real counterparts. Noteworthy scores included an average SSIM of 0.85, an average PSNR of 30 dB, and an average DSC of 0.89. The study underscores the DCGAN-MRI model's potential applications in medical imaging, including data augmentation for brain tumor detection algorithms, diverse visualizations offering insights into tumor characteristics, and enhanced data privacy through synthetic data substitution.

2.4 AGGrGAN

The AGGrGAN in [7] introduces a novel GAN architecture tailored for high-fidelity synthetic MRI scans of brain tumors, employing a fusion of three GAN models: two DCGAN and a WGAN. Metrics including SSIM, PSNR, and precision/recall were employed for evaluation. AGGrGAN consistently outperformed real images across all metrics, with SSIM scores of 0.65 (Glioma), 0.57 (Meningioma), and 0.83 (Pituitary); PSNR values of 44.17, 39.80, and 61.70; and precision/recall rates for Glioma (94.5%, 93.5%), Meningioma (93.2%, 94.0%), and Pituitary (97.7%, 97.3%). Qualitative assessment affirmed indistinguishability from real images, emphasizing AGGrGAN's effectiveness in generating realistic and precise brain tumor images [6].

2.5 StyleGAN and StyleGAN2

StyleGAN, introduced in [10,11] 2018, enhances GAN performance by refining the generator architecture and introducing intermediate latent codes. StyleGAN2 builds upon this success, addressing artifacts in generated images and offering architectural improvements. Notably, Tariq et al. apply StyleGAN2 with adaptive discriminator augmentation to generate high-quality brain tumor MRI images, addressing data limitations. StyleGAN3 further advances style control and image quality, providing a flexible framework for diverse and lifelike image synthesis [12,13].

Generator Architecture. StyleGAN and StyleGAN2 employ a progressive growing strategy, training the generator with convolutional, normalization, and upsampling layers. They introduce noise into the input for improved style mixing and use intermediate latent spaces for disentangling variational factors within the network [10,11].

Discriminator Architecture. The discriminator in StyleGAN and StyleGAN2 distinguishes real from synthetic images using convolutional, normalization, and downsampling layers [10,11].

2.6 Transfer Learning

Transfer learning involves repurposing knowledge from one domain to enhance performance in a related domain. It has been crucial in addressing data limitations and improving brain tumor detection.

3 Methodology

Despite the remarkable success of GANs in generating synthetic data, challenges persist in ensuring that the generated images are not only visually realistic but also clinically meaningful. Specifically, the challenge lies in capturing the intricate details and characteristics of brain tumor images, including tumor morphology, texture, and context. Our research aims to leverage the power of StyleGAN3 to create synthetic brain tumor images that exhibit high fidelity and realism, while also adhering to the complex medical characteristics present in real-world data.

3.1   StyleGAN3 Architecture

StyleGAN3 is an advanced generative model that extends the successes of its predecessors, including StyleGAN and StyleGAN2. It introduces innovative architectural components and training strategies to improve the quality, diversity, and control of generated images.

Generator and AdaIN Layer. StyleGAN3's generator is a deep CNN that employs a progressive growing strategy during training. This approach enables the generator to learn image synthesis at varying resolutions, progressively refining image details as training advances. It comprises a series of convolutional layers, normalization layers, and upsampling layers.

A significant innovation in StyleGAN3 is the integration of the AdaIN layer. AdaIN dynamically adjusts the mean and variance of each feature map, providing greater control over the style and appearance of the generated images. It is defined as:

$$\text{AdaIN}(F, \mathbf{y}) = \mathbf{y}_{\text{scale}} \left(\frac{F - \mu(F)}{\sigma(F)} \right) + \mathbf{y}_{\text{bias}} \tag{1}$$

Here: - $\mathbf{y}_{\text{scale}}$ and $\mathbf{y}_{\text{bias}}$ are learnable scale and bias parameters. - $\mu(F)$ and $\sigma(F)$ represent the mean and standard deviation of the feature map F.

This dynamic adjustment of style parameters contributes to the generation of diverse and high-quality images with fine-grained details.

Discriminator Architecture and PatchGAN Layer. StyleGAN3's discriminator, like the generator, is a deep CNN designed to distinguish between real and synthetic images. It features convolutional layers, normalization layers, and downsampling layers. A notable addition in StyleGAN3 is the PatchGAN layer within the discriminator, which focuses on smaller image patches. This architecture emphasizes local image details and enhances the discriminator's ability to detect synthetic images.

The combination of these architectural elements in StyleGAN3 results in the generation of high-quality, diverse, and realistic images. StyleGAN3 represents a significant advancement in the field of generative image synthesis, facilitating applications in various domains, including medical image generation [14] (Fig. 1).

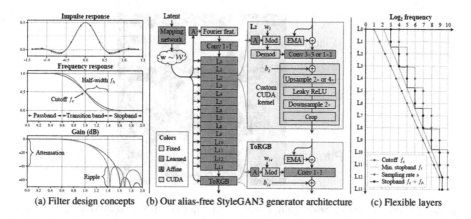

(a) Filter design concepts   (b) Our alias-free StyleGAN3 generator architecture   (c) Flexible layers

Fig. 1. Architecture of StyleGAN3 generator [14].

3.2   Integration of Components

The components of the StyleGAN3 architecture, including the progressive growing strategy, AdaIN layers, PatchGAN discriminator, and multi-component loss
function, work synergistically to produce high-quality, diverse, and realistic
images. By carefully integrating these components, StyleGAN3 achieves remarkable improvements in image synthesis over its predecessors.

The StyleGAN3 architecture represents a significant advancement in generative image synthesis. Its innovative features, such as AdaIN layers and PatchGAN
discriminator, contribute to improved style control, local detail preservation, and
stable training. StyleGAN3's architecture serves as a foundation for generating
high-quality images in various creative and practical applications [14] (Fig. 2).

3.3   Dataset

We applied the models to the brain tumor dataset as available via Kaggle.
This dataset includes 155 brain MRI samples and contains 3064 T1-weighted
images with high contrast consisting of three kinds of brain tumors which are
classified as Glioma, Meningioma, and Pituitary Tumor [15].

3.4   Dataset Preprocessing

The raw brain tumor dataset was collected and curated for training the Style-
GAN3 model. To ensure compatibility with the model architecture, the images
were uniformly resized to a resolution of 512 x 512 pixels. Additionally, to con-
form to the RGB image format required by StyleGAN3, grayscale images were
converted to the RGB color space.

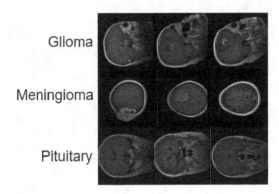

Fig. 2. Sample images in brain tumor dataset34 where each row represents images of the same class [15]

To optimize data loading and model convergence, all images were transformed into PNG format. This format is compatible with the image ingestion process of StyleGAN3, allowing smooth data input during training.

3.5   Model Configuration

For transfer learning, we employed the pre-trained afhq512 model. This model was fine-tuned to generate synthetic brain tumor images. The network architecture, consisting of the generator and discriminator, was retained from the StyleGAN3 framework. The generator incorporated the novel AdaIN (Adaptive Instance Normalization) layer for enhanced style control and diversity in the generated images.

3.6   Experimental Investigations: Hyperparameter Impact on StyleGAN3 Synthesis

The training process involved the utilization of a resized and preprocessed brain tumor dataset. To comprehensively evaluate the impact of various hyperparameters on image generation, we conducted three distinct experiments, systematically manipulating key parameters. Specifically, we focused on the gamma value and the size of the training dataset to assess their influence on the model's behavior and performance.

In our experiments, we sought to uncover the nuanced effects of gamma values on the generated brain tumor images. By systematically varying the gamma parameter, we aimed to observe how adjustments in contrast and brightness influence the realism and clinical fidelity of the synthetic dataset. Additionally, we explored the impact of altering the number of images in the training dataset, acknowledging its potential role in shaping the diversity and generalization capabilities of the StyleGAN3 model.

The training iterations were executed on the NVIDIA A100-SXM GPU, harnessing its high-performance computing capabilities to efficiently update the model parameters. Each training iteration was designed to enhance the generator's proficiency in synthesizing realistic brain tumor images, aligning with the specified characteristics inherent in the StyleGAN3 architecture.

By meticulously varying these hyperparameters, our experimental design aimed to provide valuable insights into the interplay between gamma values, dataset size, and the model's ability to generate authentic brain tumor images. Subsequent sections will delve into the specific results obtained from these experiments, shedding light on the optimal configurations that balance realism and diversity in the synthetic dataset.

[1] Each training iteration aimed to improve the generator's ability to synthesize realistic brain tumor images, aligning with the desired characteristics outlined in the StyleGAN3 architecture.

3.7 Generated Image Format

The generated synthetic brain tumor images were produced in PNG format, matching the input format during training. This format ensured consistency between the training data and the generated images, facilitating further analysis and evaluation.

3.8 Experiments

In this section, we detail the experimental setup and results of our synthetic brain tumor data generation using StyleGAN3.

Experiment 1: Gamma = 6.6, Dataset = 310 Images, Batch Size = 4

For the first experiment, we set the gamma value to 6.6 in [18] and utilized a dataset consisting of 155 brain tumor images. To augment the dataset, each image was mirrored, resulting in a total of 310 training samples. The model was trained using these samples, and the generated images were evaluated based on quality and resemblance to real brain tumor images.

Experiment 2: Gamma = 8.2, Dataset = 310 Images, Batch Size = 4

In the second experiment, we increased the gamma value to 8.2 while maintaining the dataset size of 310 images (mirrored to double the effective training data). This experiment aimed to observe the impact of higher gamma values on the generated brain tumor images. We examined whether the increase in gamma influences image realism and diversity.

Experiment 3: Gamma = 8.2, Dataset = 155 Images, Batch Size = 4

The third experiment focused on a reduced dataset size while keeping the gamma value at 8.2. We used 155 brain tumor images (mirrored to 310 training

---

[1] The code for this project can be found on the GitHub repository[2].

samples) to investigate how the model performs with limited training data. This experiment aimed to assess the model's ability to generate meaningful synthetic brain tumor images under data scarcity conditions.

3.9 Evaluation Metrics

The quality of the generated images from each experiment was evaluated using the FID [16] and KID [17] metrics. FID measures the similarity between the distribution of real and generated images, while KID assesses the difference in feature representations. Lower FID and KID scores indicate higher image quality and similarity to real data.

4 Results and Discussions

In this section, we present a detailed analysis of the results obtained from our experiments on generating synthetic brain tumor data using StyleGAN3. We conducted three experiments, each with different parameter settings, and evaluated the quality of the generated data using FID and KID metrics.

The FID and KID are metrics commonly used to assess the quality of generative models like StyleGAN3. FID measures the distance between the distributions of real and generated data. It involves extracting features using the Inception v3 network, calculating mean and covariance, and determining the Mahalanobis distance. A lower FID indicates greater similarity between the distributions, implying that generated brain tumor data closely resembles real tumors in shapes, sizes, and textures.

Similarly, KID assesses distribution similarity using the Kullback-Leibler divergence (KL divergence) instead of the Mahalanobis distance. It estimates probability density functions using Gaussian kernels and calculates KL divergence between them. A lower KID score signifies a closer match between the distributions, indicating that synthetic brain tumor data aligns well with real tumor characteristics. We use an interval of 10 ticks in every experiment, representing training progression on 40 k images.

4.1 Experiment 1: Gamma = 6.6, Dataset = 310 Images, Batch Size = 4

For the first experiment, we set the gamma value to 6.6 and used a dataset of 310 brain tumor images (mirrored for augmentation). The evaluation results for Experiment 1 are as follows:

- FID Scores: [283.376, 66.928, 67.078, 83.045, 131.257]
- KID Scores: [0.238, 0.023, 0.012, 0.024, 0.134]

The FID and KID scores provide insights into the dissimilarity and distribution similarity between the generated synthetic brain tumor images and real images from the dataset. It is worth noting that the increase in FID and KID scores may be attributed to the lower gamma value in this experiment. The choice of gamma influences the diversity and complexity of the generated images. While a lower gamma value may lead to higher diversity, it might also result in images that deviate more from the true distribution.

4.2   Experiment 2: Gamma = 8.2, Dataset = 310 Images, Batch Size = 4

Motivated by the findings from Experiment 1, we conducted a second experiment with an increased gamma value of 8.2 while maintaining a dataset size of 310 images (mirrored). The evaluation results for Experiment 2 are as follows:

- FID Scores: [283.243, 66.238, 65.241, 67.893, 67.856]
- KID Scores: [0.238, 0.023, 0.012, 0.014, 0.013]

It appears that increasing the gamma value has resulted in better FID and KID scores. This means that the generated images are now more similar to the actual data distribution, which is a positive development. This finding suggests that using a higher gamma value can potentially improve the quality of synthetic brain tumor images, making them more realistic and closer to actual medical images.

4.3   Experiment 3: Gamma = 8.2, Dataset = 155 Images, Batch Size = 4

To explore the impact of dataset size on the generated data, we conducted a third experiment using a reduced dataset of 155 images (mirrored) while maintaining the gamma value of 8.2. The evaluation results for Experiment 3 are as follows:

- FID Scores: [283.215, 73.249, 73.490, 76.187, 76.275]
- KID Scores: [0.238, 0.018, 0.021, 0.022, 0.021]

It was observed that a decrease in the size of the dataset resulted in a slight increase in FID scores and a minor variation in KID scores. This suggests that while a smaller dataset may still produce reasonable results, training with a larger dataset can lead to more stable and accurate training, resulting in lower FID and KID scores.

4.4   Visual Results

See Figs. 3, 4, and 5.

Fig. 3. Synthetic Brain Tumor MRI images for Experiment 1: Gamma = 6.6, Dataset = 310 Images, Batch Size = 4.

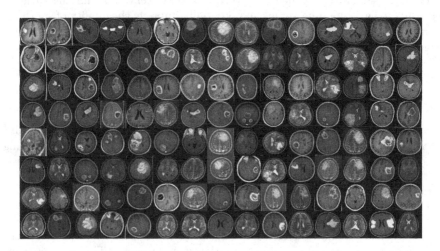

Fig. 4. Synthetic Brain Tumor MRI images for Experiment 2: Gamma = 8.2, Dataset = 310 Images, Batch Size = 4.

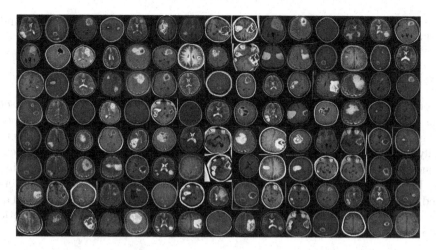

Fig. 5. Synthetic Brain Tumor MRI images for Experiment 3: Gamma = 8.2, Dataset = 155 Images, Batch Size = 4.

4.5   Discussion of Results

The analysis of FID and KID scores in different experiments offers valuable insights into the relationship between gamma values and dataset sizes in generating synthetic brain tumor data. Experiment 2's positive outcomes, which were achieved by increasing the gamma value, highlight the importance of fine-tuning hyperparameters for optimal performance. Additionally, the comparison between Experiment 1 and Experiment 3 emphasizes the significance of dataset size in generating high-quality synthetic data.

Overall, our findings indicate that StyleGAN3 shows promise in generating synthetic brain tumor data, with the potential to adapt to different scenarios and requirements. Further investigations could explore the impact of other hyperparameters and training strategies to enhance the generated data's quality and realism (Table 1).

Table 1. Comparison of final FID and KID values for each experiment.

Experiment	FID	KID
1	131.25	0.134
2	67.85	0.0138
3	76.27	0.021

5   Future Work

While our current experiments have provided valuable insights into generating synthetic brain tumor data using StyleGAN3, there are several directions for future research and enhancements that could lead to even more accurate and realistic results.

5.1 Regularization Techniques

In our experiments, we focused on manipulating the gamma value and dataset size as a means to control the diversity and quality of generated brain tumor images. However, incorporating advanced regularization techniques can potentially further improve the stability and convergence of the training process. Techniques such as spectral normalization, gradient penalty, and feature matching can be explored to mitigate mode collapse and improve the diversity of generated images. Regularization methods can also aid in preventing the generation of unrealistic or implausible images, enhancing the overall quality of the synthetic data.

5.2 Semi-supervised Learning

Utilizing the generated synthetic data for semi-supervised learning tasks is another avenue for future exploration. By integrating the synthetic data with a limited amount of real labeled data, semi-supervised learning can potentially improve the generalization and performance of models trained on limited labeled data. Investigating the effectiveness of semi-supervised learning on brain tumor image segmentation, classification, or detection tasks could be a promising research direction.

5.3 Conclusion

In conclusion, our experiments lay the foundation for further advancements in generating high-quality synthetic brain tumor data. By exploring regularization techniques, multi-modal synthesis, domain adaptation, semi-supervised learning, and ethical considerations, future research can unlock new possibilities for enhancing medical image analysis and facilitating advancements in brain tumor diagnosis and treatment.

References

1. Brock, A., Donahue, J., Simonyan, K.: Synthesizing images with style-based generative adversarial networks. arXiv preprintarXiv:1912.04994 (2019)
2. Cirillo, M., Abramian, D., Eklund, A.: Generating synthetic brain tumor images using conditional generative adversarial networks. Med. Image Anal. **64**, 101756 (2020)
3. Nalepa, J., Marcinkiewicz, M., Kawulok, M.: Data augmentation for brain tumor segmentation using StyleGAN. Front. Comput. Neurosci. **15**, 682 (2021)
4. Barocas, A., Selbst, A. D.: Fairness in machine learning. arXiv preprintarXiv:1607.05223 (2016)

5. Amodei, D., Kostyuk, N.: Accelerating deep learning research through cloud-based benchmarking. arXiv preprintarXiv:1907.00245 (2019)
6. Mukherkjee, D., Saha, P., Kaplun, D., et al.: Brain tumor image generation using an aggregation of GAN models with style transfer. Sci. Rep. **12**, 9141 (2022). https://doi.org/10.1038/s41598-022-12646-y
7. Wang, J., Li, Q., Zhang, X., Liu, H., Liu, Y.: AGGrGAN for brain tumor MRI generation using an aggregation of GANs. Sci. Rep. **12**(1), 10629 (2022). https://doi.org/10.1038/s41598-022-13060-4
8. Radford, A., Metz, L., Chintala, S.: Unsupervised representation learning with deep convolutional generative adversarial networks. arXiv preprint arXiv:1511.06434 (2015)
9. Islam, J., Zhang, Y.: GAN-based synthetic brain PET image generation. Brain Inform. **7**(1), 1–12 (2020). https://doi.org/10.1186/s40708-020-00104-2
10. Karras, T., Laine, S., Aittala, M., Hellsten, J., Lehtinen, J., Aila, T.: Analyzing and improving the image quality of StyleGAN. In: IEEE CVPR, pp. 8107–8116 (2020)
11. Karras, T., Laine, S., Aila, T.: A style-based generator architecture for generative adversarial networks. In: IEEE CVPR, pp. 4396–4405 (2019)
12. Usama, T., et al.: Brain tumor synthetic data generation with adaptive StyleGANs. In: Longo, L., ÓReilly, R. (eds.) Artificial Intelligence and Cognitive Science. AICS 2022. Communications in Computer and Information Science, vol. 1662, pp. 12–20. Springer, Cham (2023). https://doi.org/10.1007/978-3-031-26438-2_12
13. Choi, Y., Uh, Y., Yoo, J., Ha, J. W.: StarGAN v2: diverse image synthesis for multiple domains. In: IEEE CVPR, pp. 8185–8194 (2020)
14. Karras, T., et al.: Alias-free generative adversarial networks. In: Presented at the NeurIPS Conference (2021)
15. Chakrabarty, N.: Brain tumor dataset. Kaggle
16. Nunn, E.J., Khadivi, P., Samavi, S.: Compound fréchet inception distance for quality assessment of Gan created images. arXiv [cs.CV] (2021)
17. Knop, S., Mazur, M., Spurek, P., Tabor, J., Podolak, I.: Generative models with kernel distance in data space. Neurocomputing **487**, 119–129 (2022)
18. Salimans, T., Goodfellow, I.J., Yuan, W., Chen, C., LeCun, Y.: Improved techniques for training GANs. In: Advances in Neural Information Processing Systems, pp. 2234-2242 (2016)

Optimized Machine Learning Models for Hepatitis C Prediction: Leveraging Optuna for Hyperparameter Tuning and Streamlit for Model Deployment

Uriel Nguefack Yefou[1]([✉]) [iD], Pauline Ornela Megne Choudja[2] [iD], Binta Sow[2] [iD], and Abduljaleel Adejumo[2] [iD]

[1] African Institute for Mathematical Sciences Cameroon, Crystal Gardens Limbe 608 South West, Limbe, Cameroon
uriel.nguefack@aims-cameroon.org
[2] African Institute for Mathematical Sciences Senegal, Km2 route de Joal (IRD Center), Mbour 1418 Thies,Mbour, Senegal
{pauline.o.m.choudja,binta.sow,abduljaleel.adejumo}@aims-senegal.org

Abstract. Machine Learning techniques have gained significant attention for their potential to solve diverse real-world problems across various fields. This study uses machine learning algorithms to predict hepatitis C stages, a prevalent liver disease affecting a substantial portion of the global population. By employing a dataset encompassing 615 patients and incorporating a multitude of factors associated with hepatitis C, a comprehensive analysis was conducted to compare the performance of six prominent machine learning algorithms. The algorithms considered include categorical boosting (CatBoost), Gaussian Naive Bayes (GNB), Random Forest (RF), Extreme Gradient Boosting (XGBoost), Light Gradient Boosting Machine (LGBM), and ExtraTreeClassifier (ExtraT). To optimize the performance of these models, a hyperparameter optimization technique called Optuna was utilized to find the ideal parameters for each algorithm. Subsequently, all models' performance was evaluated using the test dataset, comprising 20% of the overall patient data. The research findings revealed that the XGBoost algorithm emerged as the most effective approach, exhibiting a remarkable accuracy of 94.31%. Furthermore, the XGBoost model demonstrated exceptional F1-score, precision, and recall values, measuring 94.23%, 94.63%, and 94.31%, respectively. Building upon these promising results, we deployed the XGBoost model in a user-friendly web application leveraging Streamlit. This deployment ensures easy accessibility and usability of the model for the broader community.

Keywords: Machine Learning · hepatitis C · Optuna · Streamlit · XGBoost · Hyperparameter

1 Introduction

The liver is identified to be one of the most vital organs with multiple functions, including secretion of bile, blood filtration, transformation, and storage of substances absorbed from the digestive system [13]. Infections affecting the liver can have detrimental consequences, potentially leading to the development of liver cancer. The hepatitis C Virus (HCV) is the cause of hepatitis C, which is an inflammatory liver disease. It is distinguished by a slow advancement of hepatic fibrosis, progressing from stage no fibrosis (stage 0) to cirrhosis (stage 4) [17]. According to [10], approximately 30% of people infected with HCV naturally remove the virus after six months without therapy, while the remaining 70% acquire chronic HCV infection, which can progress to cirrhosis and liver cancer within 20 years. According to the World Health Organisation (WHO), HCV is a global health issue, with an estimated 71 million individuals living with chronic HCV infection in 2015 and over 1.75 million new infections emerging each year. The disease is prevalent across regions including the Americas, Europe, Asia, Africa, etc. [21]. Unfortunately, vaccines for hepatitis C have not been discovered [27]. The diagnosis of HCV infection involves two steps. Firstly, a serological test is conducted to detect anti-HCV antibodies, identifying infected individuals. In addition, patients who test positive for the antibody are given a nucleic acid test to determine whether they have a persistent infection with HCV and whether they require therapy [22]. Upon diagnosis of chronic HCV infection, an evaluation is performed to determine the extent of liver damage in terms of fibrosis and cirrhosis. However, this diagnostic approach is time-consuming, expensive, and limited in resource-constrained settings.

Early detection plays a pivotal role in mitigating the potential health consequences associated with infection and curbing the transmission of the virus. Leveraging the advancements in technology, the utilization of machine learning (ML) algorithms, in conjunction with medical expertise, holds the promise of facilitating rapid and efficient automated diagnosis. ML algorithms have the capability to extract critical insights from data encompassing patient information, enabling the identification of individuals who are most likely to derive benefits from diagnostic testing and subsequent treatment interventions. This approach not only enhances patient outcomes but also contributes to the reduction of healthcare expenditures. Overall, the integration of ML algorithms in forecasting and preventing hepatitis C infection can yield substantial improvements in public health by enabling proactive interventions and curtailing the spread of the disease.

The primary aim of this work is to optimize ML models for the accurate prediction of hepatitis C. In addition, a key focus is to enhance the interpretability and transparency of these models, enabling healthcare professionals to comprehend the underlying rationale behind the predictions and validate the reliability of the models. This will empower physicians with actionable insights derived from ML algorithms, thereby facilitating informed decision-making and improving the overall effectiveness of hepatitis C diagnosis and treatment.

The paper follows this organization: In Sect. 2, a review of the pertinent literature concerning ML used in classifying hepatitis C disease is provided. Section 3 presents the methodology adopted in this research along with a comprehensive description of the dataset used. Moving on, Sect. 4 presents the study's results and an analysis of the experimental testing process. Finally, in Sect. 5, the paper is concluded, and recommendations for future research are put forth.

2   Related Work

In [16], the authors used the XGBoost algorithm to predict hepatitis C in blood donors and clinical datasets from chronic hepatitis C patients. After evaluation of the model, the experimental results indicate that the XGBoost model is robust and performs better with an accuracy of 91.56% compared with the results obtained from Support Vector Machine (SVM), K-Nearest Neighbours (KNN), Decision Tree (DT), and Adaboost algorithms. Similarly, [20] conducted a study using four (04) ML techniques namely KNN, SVM, Naive Bayes, and DT for the classification, prediction, and diagnosis of hepatitis C. Based on the analysis of 615 records, the DT method exhibited superior performance, achieving an accuracy of 93.44%, outperforming the other models in terms of the specific classification objectives.

[18], and [25] conducted a study in which they developed an HCV predicting model with ML techniques, specifically KNN classifiers, and RF. The dataset used in their study was sourced from the UCI-ML repository and consisted of 668 instances of HCV. The result indicated that the RF model gave the best accuracy score of 94.88%.

In [19], the authors addressed the imbalance problem in the dataset obtained from the UCI-ML Repository by applying the technique SMOTE (Synthetic Minority Oversampling Technique). Their results show that RF performed best across various evaluation metrics. Employing the XGBoost model, [7] and [2] predicted hepatitis C disease with an accuracy of less than 90%.

[3] used ensemble-based ML models to identify whether cirrhosis is present in patients with hepatitis C. Four ML models, Gradient Boosting Machine, RF, ExtraT, and XGBoost were trained on the data, composed of 28 features with a total of 2038 patients from Egypt. The result shows that the ExtraT model performed better than the other two models with an accuracy of 96.92%, a precision of 99.81%, and a recall of 94.00% using 16 features of the 28.

3   Methods

3.1   Dataset Description

The Intelligent Systems and Machine Learning Center at the California University, Irvine (UCI) provided the data for the UCI dataset, which was generated by [15] and contains information from 615 individuals. The target feature, bilirubin (BIL), gender, age, blood levels of ALB (Albumin), ALP (Alkaline

Phosphatase), ALT (ALanine aminoTransferase), CHE (Choline Esterase), AST (ASpartate aminotransferase), GGT (Gamma-Glutamyl-Transferase), CHOL (CHOLesterol), CREA (CREAtinine Blood test) and PROT (Total protein test) are among the thirteen features that make up each individual's record. These features are used to classify each person as either a blood donor or as having hepatitis C disease, which includes its development to cirrhosis and fibrosis. The ML algorithms that will be used to determine the possibility of a person contracting the virus will be trained and tested on this data.

Using Pearson's approach [9], this study determined the correlation coefficients between various features shown in Fig. 1. Pearson's correlation coefficient, which has values ranging from -1 to 1, can be used to assess the linear relationship between two variables, with -1 signifying a perfect negative correlation, 0 indicating no correlation, and 1 representing a perfect positive correlation. It assesses the degree and direction of the relationship between two variables. The connection between features in the dataset used can provide insight into how different variables affect the development of hepatitis C virus (HCV) infection. It can, for example, assess whether older people are more likely to HCV infection by examining the link between the HCV infection rate and the age.

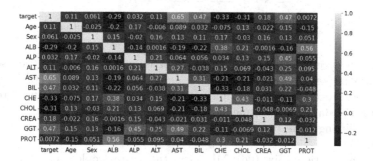

Fig. 1. Correlation plot HCV Disease.

3.2   Data Preprocessing

The initial stage in the proposed system is data preparation or preprocessing, which involves removing noisy values and replacing missing values for specific attributes. The target column 'Category' which contains 5 classes was encoded as follows: '0=blood donor' by 0, '0s=suspect blood donor' by 1, '1=hepatitis' by 2, '2=Fibrosis' by 3 and '3=cirrhosis' by 4. For the sex column, male was encoded as 1 and female as 0. The empty values were filled with 10% trim_mean of the column using the library trim_mean of the python package 'scipy.stats'. Our variables were scaled using the StandardScaler Python library of scikit-learn. Since our dataset was highly unbalanced, the 'SMOTE' [6] technique was employed to solve the imbalance problem.

3.3 Proposed Framework

The suggested framework is divided into 06 stages, including (1) Data collection, (2) Data pre-processing, (3) data splitting, (4) model generation, (5) model optimization, (6) model evaluation, and (7) deployment of the best model. Figure 2 shows the proposed framework for predicting the HCV disease.

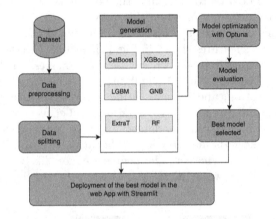

Fig. 2. Proposed framework for hepatitis C detection.

- Data collection: In this stage, our dataset is imported and will be used for the next phases.
- Data pre-processing: The techniques presented in Sect. 3.2 are used to pre-process the data before modeling.
- Data splitting: The data is split into 2 parts: test set(20%), and training set(80%).
- Model generation: Six (06) ML algorithms are implemented: Catboost, XGBoost, LGBM, GNB, ExtraT, and RF. These algorithms will be more explained in Sect. 3.4.
- Model optimization: A technique called Optuna is used to identify the best parameters for each of the implemented ML algorithms. Optuna [1] is an autonomous hyperparameter optimization software framework that is specifically built for ML.
- Model Evaluation: The models's performance is assessed using the test set for future comparison.
- Deployment of the best model: Following the evaluation phase, the best model is selected and implemented in our web application using Streamlit.

3.4 Description of Utilized ML Techniques

Ensemble algorithms combine the outputs of numerous models that have been trained. On one side, Boosting classifiers are iterative ensemble algorithms that adjust the weight of an observation based on recent classification results. On the other side, Bagging classifiers use weighted averages or majorities to merge many independent variables.

In this investigation, three boosting approaches (CatBoost, XGBoost, and LGBM), a straightforward ML algorithm (Gaussian Naive Bayes), and two bagging methods (ExtraT and RF) were employed.

1. **CatBoost**

   When utilized for Supervised ML problems, CatBoost, a component of the Gradient Boosted Decision Tree (GBDT) ensemble machine learning approaches, introduces two novelties: Ordered Boosting and Ordered Target Statistics [12]. It is an innovative algorithm used to process categorical features [23].

2. **XGBoost**

   Extreme Gradient Boosting is a ML approach built on Decision Trees. XGBoost addresses overfitting by including additional regularization in its goal function, which can be an issue for ensemble models. This regularization component penalizes the model's complexity, improving generalizability and lowering the risk of overfitting [8]. The prediction of the model is given by:

   $$\hat{y}_i = \sum_{k=1}^{K} f_k(x_i), \qquad f_k \in \mathcal{F} \tag{1}$$

   where $f_k(x_i)$ is the prediction of the $k - th$ tree, $\mathcal{F}$ is the space of Regression Tree and K the total number of trees.

3. **LGBM**

   LGBM, or Light Gradient Boosting Model is used when there are more variables in the data, which leads to a problem of scalability and effectiveness of the model. The primary reason for this behavior is that it takes a significant amount of time and effort for each feature to scan through all of the different data examples and compute all of the probable split points. It uses two techniques: the Gradient-based On-Side Sampling (GOSS) and the Exclusive Feature Bundling (EFB). In reality, GOSS will just utilize the remaining data to compute the total information gain, removing a large portion of the data section with weak gradients. In order to decrease the number of features, the EFB typically takes no non-zero value at the same time as the mutually exclusive features [14].

4. **Gaussian Naive Bayes**

   Gaussian Naive Bayes(GNB) is a sophisticated ML technique for classification tasks that utilizes the rules of Bayes and supposes that each class follows a Gaussian distribution. Low variance, Incremental learning, Computational efficiency, Robustness in the face of absent value, and Robustness in facing noise are important features of Gaussian Naive Bayes [4].

5. **Random Forest**

Random Forest (RF) [5] is an approach to classification and regression using ensemble learning. RF generates multiple decision trees by employing random subsets of both the features and the training data during the construction process. By averaging or tallying all of the different trees' forecasts, the ultimate prediction is made.

6. **ExtraT**

The Extra Tree [11], also known as the Extremely Randomized Tree, generates predictive models for classification and regression problems. It's comparable to other algorithms like Decision Trees and Random Forests, but it provides superior predictions by including additional data facts.

Additionally, the extra tree technique is faster and easier to use than earlier approaches. Therefore, it is a powerful data mining and predictive modeling tool.

3.5   Performance Evaluation

The result of the classification evaluation in terms of *accuracy, recall, F1 score,* and *precision* are well-known evaluation measures that can be used to assess the performance of the proposed approach. These measures have been developed based on a confusion matrix containing FNs (False Negatives), FPs (False Positives), TPs (True Positives), and TNs (True Negatives) [24,26].

$$Accuracy = \frac{TP + TN}{TP + TN + FP + FN} \tag{2}$$

$$Precision = \frac{TP}{TP + FP} \tag{3}$$

$$Recall = \frac{TP}{TP + FN} \tag{4}$$

$$F1\ score = 2 \times \frac{Precision \times Recall}{Precision + Recall} \tag{5}$$

3.6   Streamlit Framework

Streamlit lets you transform data scripts into shareable web applications in just a few minutes. It's a free, open-source Python application. Once you've created an application, you can use the Community Cloud platform to deploy, manage and share it. The best model selected from the set of all our models is deployed using Streamlit. The model can be tested by all users of the application.

4 Results

This section is organized into four distinct parts: the first addresses the effectiveness of the optimization technique on the different models developed. The second part will focus on the important feature of the best model, the XGBoost model. The confusion matrix of the XGBoost model on test data will be provided in the third section, and the Web App developed with Streamlit will be presented in the last section.

4.1 Model Comparison

The obtained results from the different models are evaluated based on accuracy, precision, recall, and the F1 score. Table 1 provides a comparison of the six models when using their default parameters. It can be observed that the CatBoost model outperforms the other models in terms of recall, accuracy, and F1 score. In contrast, when considering precision as the primary evaluation metric, the Gaussian Naive Bayes (GNB) model demonstrates superior performance.

Table 1. Performance of the model without optimization.

Models	Accuracy	Precision	Recall	F1-score
CatBoost	**0.9186**	0.9155	**0.9187**	**0.9115**
XGBoost	0.9024	0.8979	0.9024	0.8973
LGBM	0.9024	0.9069	0.9024	0.8954
GNB	0.9106	**0.9169**	0.9106	0.9089
ExtraT	0.8780	0.8763	0.8780	0.8498
RF	0.8862	0.8889	0.8862	0.8687

Optuna framework was employed to perform a hyperparameter optimization process for each model. The best parameters were subsequently used to train the models and the outcomes are reported in Table 2. This table indicates that the XGBoost model outperforms the other models across all evaluated metrics and its best hyperparameter values are shown in Sect. 6.

Table 2. Performance of the models with optimization.

Models	Accuracy	Precision	Recall	F1-score
CatBoost	0.9430	0.9434	0.9431	0.9409
XGBoost	**0.9431**	**0.9463**	**0.9431**	**0.9423**
LGBM	0.9350	0.9337	0.9350	0.9304
GNB	0.9106	0.9169	0.9106	0.9089
ExtraT	0.9268	0.9300	0.9268	0.9236
RF	0.8943	0.9124	0.8943	0.8854

4.2   Feature Importance on the XGBoost Model

Feature importance plot gives the visual representation of the features that contribute most to the model. Figure 3 shows that all the features had a significant contribution to the model, given data no weight is less than 0.04.

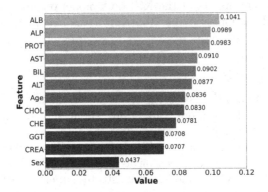

Fig. 3. XGBoost Feature Importance.

The ALB had the highest contribution in Fig. 3 followed by the ALP and PROT respectively. Sex had the least contribution to the XGBoost model.

4.3   Confusion Matrix Plot

Figure 4 presents the confusion matrix obtained from the testing set using the XGBoost classifier. In this matrix, each row corresponds to the true label, representing the patients' status, while each column represents the predicted label.

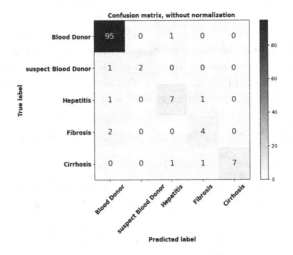

Fig. 4. XGBoost confusion matrix.

The model accurately predicted almost all Blood donor cases except one, and correctly identified 2 (out of 3) suspected blood Donors, 7 (out of 9) hepatitis cases, 4 (out of 6) Fibrosis cases, and 7 (out of 9) Cirrhosis cases.

4.4 Presentation of the Web Application

Using the XGBoost classifier, a web application was designed with Streamlit, and the result is presented in Fig. 5.

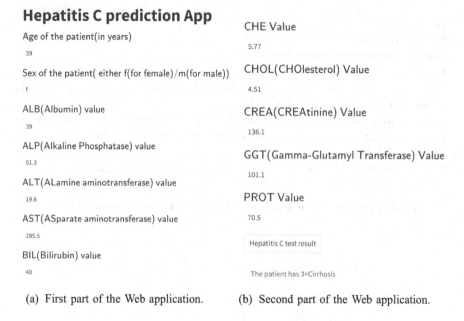

(a) First part of the Web application.   (b) Second part of the Web application.

Fig. 5. Web application realized with Streamlit.

Based on the given example, the model's prediction indicates that the patient has 'Cirrhosis'.

5   Conclusion

The integration of ML has significantly enhanced clinicians' diagnostic capabilities, leading to a notable reduction in the time required for disease diagnosis. This study proposes an ML framework based on CatBoost, XGBoost, LGBM, GNB, ExtraT, and RF algorithms, employing the Optuna optimization technique to classify and predict patients infected with HCV. Through a comprehensive comparative analysis of the models using the test set, XGBoost outperformed all other models across all evaluation metrics. Furthermore, this best model was deployed using Streamlit, which facilitated smoother and easier testing of the application. Future research should focus on augmenting the model's efficiency by incorporating additional hepatitis C-related variables. Additionally, gather more patient data to enhance the model's performance and assess the robustness of the model by evaluating it on patient data collected from different regions.

Declarations

- Funding
  This research did not receive any specific grant from funding agencies in the public, commercial, or not-for-profit sectors.
- Conflict of interest
  The authors declare no conflicts of interest.
- Consent for publication
  All authors have read and agreed to the published version of the manuscript.
- Availability of data and materials
  Publicly available datasets were analyzed in this study. This data can be found here: https://archive.ics.uci.edu/ml/datasets/HCV+data.
- Code availability
  The code is available upon request from the corresponding authors.

6   Best Hyperparameters for the XGBoost Model

(See Table 3).

Table 3. Best Hyperparameters for XGBoost

Hyperparameter	Value
max_depth	4
learning_rate	0.1969097619840202
n_estimators	197
min_child_weight	1
gamma	0.09693404552771179
subsample	0.036768173724319515
colsample_bytree	0.19587409453980753
reg_alpha	0.41354766675033994
reg_lambda	0.009420448414595634

References

1. Akiba, T., Sano, S., Yanase, T., Ohta, T., Koyama, M.: Optuna: a next-generation hyperparameter optimization framework. In: Proceedings of the 25th ACM SIGKDD International Conference on Knowledge Discovery and Data Mining, pp. 2623–2631 (2019)
2. Alizargar, A., Chang, Y.L., Tan, T.H.: Performance comparison of machine learning approaches on hepatitis c prediction employing data mining techniques. Bioengineering 10(4), 481 (2023)
3. Alotaibi, A., et al.: Explainable ensemble-based machine learning models for detecting the presence of cirrhosis in hepatitis c patients. Computation 11(6), 104 (2023)
4. Anand, M.V., KiranBala, B., Srividhya, S., Younus, M., Rahman, H., et al.: Gaussian naïve bayes algorithm: A reliable technique involved in the assortment of the segregation in cancer. Mobile Information Systems 2022 (2022)
5. Breiman, L.: Random forests. Machine Learn. 45, 5–32 (2001)
6. Chawla, N.V., Bowyer, K.W., Hall, L.O., Kegelmeyer, W.P.: Smote: synthetic minority over-sampling technique. J. Artif. Intell. Res. 16, 321–357 (2002)
7. Chen, L., Ji, P., Ma, Y.: Machine learning model for hepatitis c diagnosis customized to each patient. IEEE Access 10, 106655–106672 (2022)
8. Chen, T., Guestrin, C.: Xgboost: A scalable tree boosting system. In: Proceedings of the 22nd ACM sigkdd International Conference on Knowledge Discovery and Data Mining, pp. 785–794 (2016)
9. Cohen, I., et al.: Pearson correlation coefficient. Noise reduction in speech processing, pp. 1–4 (2009)
10. Gerber, M.A.: Pathology of hepatitis c. FEMS Microbiol. Rev. 14(3), 205–210 (1994)
11. Geurts, P., Ernst, D., Wehenkel, L.: Extremely randomized trees. Mach. Learn. 63, 3–42 (2006)
12. Hancock, J.T., Khoshgoftaar, T.M.: Catboost for big data: an interdisciplinary review. J. big data 7(1), 1–45 (2020)
13. Kalra, A., Yetiskul, E., Wehrle, C.J., Tuma, F.: Physiology, liver (2018)
14. Ke, G., et al.: Lightgbm: a highly efficient gradient boosting decision tree. In: Advances in Neural Information Processing Systems 30 (2017)
15. Lichtinghagen, R., Klawonn, F., Hoffmann, G.: Hcv data data set. Available online:(accessed on 19 March 2023), UCI Machine Learning Repository (2020)
16. Ma, L., Yang, Y., Ge, X., Wan, Y., Sang, X.: Prediction of disease progression of chronic hepatitis c based on xgboost algorithm. In: 2020 International Conference on Robots & Intelligent System (ICRIS), pp. 598–601. IEEE (2020)
17. Marcellin, P., Asselah, T., Boyer, N.: Fibrosis and disease progression in hepatitis c. Hepatology 36(S1), S47–S56 (2002)
18. Nandipati, S.C., XinYing, C., Wah, K.K.: Hepatitis c virus (hcv) prediction by machine learning techniques. Appl. Modell. Simul. 4, 89–100 (2020)
19. Oladimeji, O.O., Oladimeji, A., Olayanju, O.: Machine learning models for diagnostic classification of hepatitis c tests. Front. Health Inform. 10(1), 70 (2021)
20. Oleiwi, A.: Development of diagnostic decision making for chronic hepatitis c virus patients by various supervised predictive model. J. Adv. Res. Dyn. Control Syst. 12, 3113–3123 (10 2020)
21. Organization, W.H., et al.: Global hepatitis report 2017: World health organization. Accessed Oct 23 2020 (2017)

22. Organization, W.H., et al.: Hepatitis C rapid diagnostic tests for professional use and/or self-testing. World Health Organization (2022)

23. Prokhorenkova, L., Gusev, G., Vorobev, A., Dorogush, A.V., Gulin, A.: Catboost: unbiased boosting with categorical features. In: Advances in Neural Information Processing Systems, vol. 31 (2018)

24. Raschka, S.: An overview of general performance metrics of binary classifier systems (2014)

25. Safdari, R., Deghatipour, A., Gholamzadeh, M., Maghooli, K.: Applying data mining techniques to classify patients with suspected hepatitis c virus infection. Intell. Med. **2**(04), 193–198 (2022)

26. Sokolova, M., Japkowicz, N., Szpakowicz, S.: Beyond Accuracy, F-Score and ROC: A Family of Discriminant Measures for Performance Evaluation. In: Sattar, A., Kang, B. (eds.) AI 2006: Advances in Artificial Intelligence, pp. 1015–1021. Springer Berlin Heidelberg, Berlin, Heidelberg (2006). https://doi.org/10.1007/11941439_114

27. Zingaretti, C., De Francesco, R., Abrignani, S.: Why is it so difficult to develop a hepatitis c virus preventive vaccine? Clin. Microbiol. Infect. **20**, 103–109 (2014)

Explainable Rhythm-Based Heart Disease Detection from ECG Signals

Dereje Degeffa Demissie[1,2]($\boxtimes$) (iD) and Fitsum Assamnew Andargie[1] (iD)

[1] School of Electrical and Computer Engineering, Addis Ababa Institute
of Technology, Addis Ababa University, Addis Ababa 385, Ethiopia
derara12@gmail.com, fitsum.assamnew@aait.edu.et
[2] Ethiopian Artificial Intelligence Institute, P.O. Box 40782, Addis Ababa, Ethiopia

Abstract. Healthcare decision support systems must operate with confidence and trust. Numerous researchers have attempted to automate the identification and classification of cardiovascular conditions from electrocardiogram (ECG) signals. One particular area of research involves utilizing deep learning (DL) for the classification of ECG signals into various heart disease classes. However, DL models do not provide information on why they reached their final decision. This makes it difficult to trust their output in a medical environment. To address this trust issue, ongoing research aims to explain the decisions made by DL models. Some approaches have been used to improve the interpretability of DL models, using the Shapley Value (SHAP). However, SHAP's explanation happens to be computationally expensive. In this research, we develop a deep learning model that can detect five rhythm-based heart diseases with explainability. We employ visual explanations algorithms, that are Grad-CAM and Grad-CAM++, as an explainability framework. These explainers are relatively lightweight and can be executed quickly on a standard CPU or GPU. Our model was trained using 12-lead ECG signals from the PTB-XL large dataset. Our model was effective, with a classification accuracy of 0.96 and an F1 of 0.88. The outputs of the model were visually explained using Grad-CAM and Grad-CAM++ of which Grad-CAM++ produced more localized explanations. To evaluate the explainability, we gave ten randomly selected outputs to two domain experts. The explanations that were not equally accepted by the experts still had a consensus in many of the individual leads out of the 12.

Keywords: Heart Disease · Rhythm-based · Explainability · ECG · Grad-CAM · Grad-CAM++

1 Introduction

Cardiovascular disease (CVD) is a disease that affects the heart and blood vessels [7]. It is currently the leading cause of death in the globe [12]. The number of CVD-related deaths has been steadily increasing, reaching 17.9 million in 2015, and it is projected to rise to 22 million by 2030 if no action is taken [7,10].

T. G. Debelee et al. (Eds.): PanAfriConAI 2023, CCIS 2068, pp. 101–116, 2024.
https://doi.org/10.1007/978-3-031-57624-9_6

As shown in Fig. 1, the ECG signal in the normal case representation has waves, intervals, segments, and one QRS complex [6]. Each represents a specific electrical event. The P wave in the ECG signal denotes the impulse produced by the Sinoatrial Node (SA) node. The contraction of the ventricles makes the QRS complex. The T wave shows repolarization of the ventricles. The RR interval helps as an illustration of heart rate irregularity [3].

As ECG is a commonly used diagnostic tool for assessing cardiovascular conditions, many studies have been done to automate the identification and classification of cardiovascular conditions from its signals using DL models [8, 10, 17]. However, DL models don't explain how they reached their final decision.

For this study, we present a modified ResNet-18 model in light of rhythm-based heart disease classification using the PTB-XL dataset of ECG signals for five rhythm-based cardiovascular diseases: Atrial fibrillation (AFIB), Sinus tachycardia (STACH), Sinus arrhythmia (SARRK), Sinus bradycardia (SBRAD), and Normal atrial pacemaker (PACE). Then solve the black-box nature of DL by using an explainability framework from a visual explanation method called gradient weighted class activation mapping (Grad-Cam) and gradient-weighted class activation mapping plus-plus (Grad-Cam++) to highlight regions of the ECG signals that were important for the classification. The list below provides our contributions

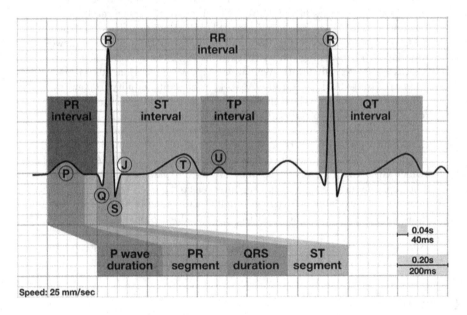

Fig. 1. ECG intervals, waves, and segments [6]

1. **Modification of the Resnet-18 model**: We adapted the Resnet-18 model architecture, which was originally designed for 2D image processing, to effectively handle 1D ECG signals. The model was modified to take into account the unique features of ECG signals. To reduce the number of model parameters, prevent overfitting, and increase computing performance, we utilized global average pooling (GAP). Each feature map of the GAP was concatenated in order to know the role of each lead in the classification of different classes.

2. **Increasing trustworthiness**: In medical applications, obtaining the trustworthiness of predictions made by DL models is important. To address this, we used the Grad-CAM++ algorithm, which generates heatmaps to visualize the important regions of ECG signals. These heatmaps provided an explanation for the classification made by our model, helping to increase the explainability and clearness of the results.

3. **Model Explainability Evaluation**: To measure the level of explainability of our deep learning model, two domain experts were consulted for their comments. They were presented with a randomly selected sample of 10 ECG records classified and highlighted with our proposed model. This evaluation allowed us to validate the explainability of our model from the perspective of experts in the field, ensuring that the explanations provided by the model were meaningful and accurate.

The remaining portion of this paper is structured as follows: Part II provides overviews of the theoretical background and related works. Section 3 introduces the proposed method, including the suggested model and the hyper parameters utilized. In Sect. 4, we present the results obtained from conducting experiments using the proposed model, as well as the explainability framework. Additionally, we include some sample results for evaluation and discuss the performance comparison. Section 5 summarizes the conclusions drawn from the study and lastly, outlines potential further research.

2 Background and Related Works

2.1 Gradient-Based Localization, Aimed at Visual Explanations

Visual explanations play a crucial role in understanding the predictions made by deep neural networks (CNNs) and increasing trust in their decision-making process. Explainability aims to open the black box of deep learning models and provide insights into how predictions are made. By explaining the rationale behind the model's decisions, users can gain a better understanding of the network's internal mechanisms. This becomes particularly important for risk-sensitive applications like clinical decision assistance. Post-hoc explanation techniques like CAM, Grad-Cam, and Grad-CAM++ are employed to explain the model's behavior during the evaluation phase.

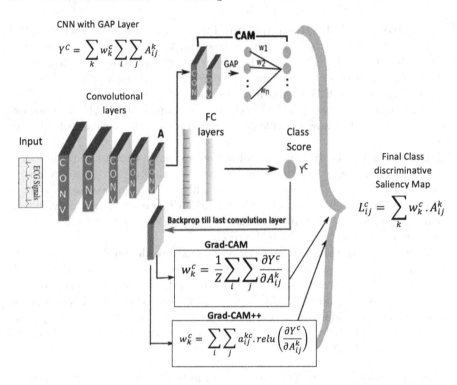

Fig. 2. A summary of each of the three techniques: CAM, Grad-CAM, and Grad-CAM++, with their corresponding computation expressions [2].

Class Activation Map (CAM): The goal of CAM is to explain how a model learns from data or why it performs badly on some tasks [14]. It is based on global average pooling (GAP), where the mean value of every vector is computed and translated into a category label. The averaged feature maps from the last convolutional layer are fed into the last fully connected layer, which is in charge of classification. A weighted sum of the feature maps for each class is used to create class activation maps. The specific saliency map for a class correlates with the significance of spatial locations in predicting that class.

$$Y^c = \sum_k W_k^c \sum_i \sum_j A_{ij}^k \qquad (1)$$

In Eq. 1, Y^c is the final classification score for a specific class c, W_k^c is the weight connected to class c and, feature map k, A^k is the last convolutional layer activation value of feature map k.

$$L_{ij}^c = \sum_k W_k^c A_{ij}^k \qquad (2)$$

Eqation 2 shows the termination of the saliency map L^c using W_k^c and A^k, which highlight the regions of the input that contribute to the model's decision-making.

Grad-CAM: Grad-CAM is an extension of CAM that considers both the gradients and weights flowing into the last convolutional layer [15]. This modification allows layers preceding the last layer to contribute to the activation map. Grad-CAM enhances the transparency of CNN-based models by visualizing input regions crucial for predictions, providing high-resolution details. It can be applied to any layer of the network, with the last layer being particularly important for identifying the signal components contributing most to the final prediction. The gradients and weights are used to compute neuron significance weights, which are then utilized to generate the Grad-CAM map.

$$W_k^c = \frac{1}{Z} \sum_i \sum_j \frac{\partial y^c}{\partial A_{ij}^k} \tag{3}$$

where Z is a fixed number (number of pixels in the activation map). In Eq. 3, $\frac{1}{Z} \sum_i \sum_j$ is used to represent the global average pooling and $\frac{\partial y^c}{\partial A_{ij}^k}$ denotes the back propagation gradients. The kth feature map is represented by A^k in the gradient expression, and y^c is the class c score. Next, the following formula yields the Grad-CAM for a class c at point (x, y):

$$L_{Grad-Cam}^c{}^{(x,y)} = ReLU\left(\sum_k W_k^c A^{k(x,y)}\right) \tag{4}$$

where the negative values were mapped to zero using the RELU operator, which is an activation function if the input is positive, returns the value; else, it returns zero. We calculate the output of Grad-CAM for each class being analyzed, just like we did with CAM.

Grad-CAM++: is a variant of Grad-CAM that aims to provide effective visual explanations for various tasks in a human-interpretable manner. It improves upon Grad-CAM by using a weighted average of partial derivatives instead of a simple sum. The weights are determined based on class scores and activation maps computed in the last convolutional layer. Grad-CAM++ introduces a parameter α to calculate the weights, and it offers a more sophisticated backpropagation process for weight estimation.

$$\alpha_{ij}^{kc} = \frac{\frac{\alpha^2 y^c}{(\alpha A ij^k)^2}}{2\frac{\alpha^2 y^c}{(\alpha A_{ij}^k)^2} + \sum_a \sum_b A_{ab}^k \left(\frac{\alpha^3 Y^c}{(\partial A_{ij}^k)^3}\right)} \tag{5}$$

$$W_k^c = \sum_i \sum_j \alpha_{ij}^{kc} ReLU E\left(\frac{\partial y^c}{\partial A_{ij}^k}\right) \tag{6}$$

Like other CAMs, it maintains that,

$$M_{Grad-Cam++}^{c}{}^{(x,y)} = ReLUE\left(\sum_k w_k^c A^{k(x,y)}\right) \tag{7}$$

Grad-CAM++ will work in the same way as the Grad-Cam formula when $\alpha_{ij}^{kc} = 1/Z$ in Fig. 2. In this case, Grad-Cam++ is generalized to Grad-Cam.

2.2 Related Works

The field of ECG signal analysis has seen advancements in computer-based interpretation since the 1950s, with the introduction of digital conversion of analog signals [16]. Numerous Machine Learning (ML) techniques, including in particular DL approaches, have shown good performance in detecting abnormal ECG waveforms and events, improving the detection accuracy of several heart-related disorders [12].

In the domain of DL models for heart disease detection, various architectures have been explored. Recursive neural networks (RNNs), convolutional neural networks (CNNs), and long-term short-term memory (LSTM) are some of the Deep Neural Networks (DNNs) that have been used. Among these, CNNs have gained popularity due to their ability to extract local features and exhibit robustness and noise tolerance in ECG classification [12]. CNNs require minimal pre-processing and can automatically learn hierarchical patterns using stacked trainable filters, or kernels. They have been successfully applied in tasks such as natural language processing, signal analysis, and image classification. CNNs with shared-weight architectures and parallelization offer fast computation and superior performance [9].

RNNs, on the other hand, are appropriate for modeling sequential data and capturing temporal dependencies in ECG signals. With iterative updates of hidden states and memory, RNNs can efficiently process inputs of varying lengths. RNNs are best for handling time series information and sequential patterns [20] which makes them ideal for processing ECG signals.

Combining the strengths of CNNs and RNNs, Convolutional Recurrent Neural Networks (CRNNs) have been developed. CRNNs leverage CNNs for extracting local features and RNNs for summarizing extracted features over time. This architecture is particularly useful for analyzing long ECG signals with varying sequence lengths and multiple channels [5].

Some approaches have incorporated interpretable expert features, multi-level attention weights, or attribute scores to enhance the interpretability of ECG data. Research in interpretability focuses on two directions: interpreting complex deep learning models using simpler models and directly constructing interpretable deep learning models by incorporating human-understandable mechanisms [4].

Notable studies in the field include the deployment of a deep CNN on a single-lead recorder system for continuous monitoring of atrial fibrillation (AF), which achieved promising performance compared to annotated reports from an insertable cardiac monitor [18]. Another study developed a deep neural network for multi-label classification of cardiac arrhythmias from 12-lead ECG records, achieving high accuracy rates using 10-fold cross-validation and employing the Shapley Additive Explanations (SHAP) technique for model interpretation [21]. Furthermore, two deep learning models, Dense Net, and CNN, were developed for myocardial infarction (MI) detection based on ECG signals, achieving high accuracy rates and utilizing class activation mapping techniques for visualizing influential ECG leads and waveform portions [11]. Although these methods show the effectiveness of deep learning models in ECG analysis, there are still computational complexity issues with the study in [21] as stated by Ayano, Y.M. et. al [1], in addition, the runtime is exponential in terms of the number of variables to analyze Kumar.I. et.al [13], and restrictions for one particular cardiac disease, MI, in the work of [11].

3 Proposed Methodology

The methodology we followed as depicted in Fig. 3 includes dataset selection, preprocessing, data splitting, model and explainability development, and model evaluation.

3.1 Dataset

The PTB-XL dataset, which is known as the largest 12-lead clinical ECG waveform dataset that is publicly accessible [19], is used as the basis for our proposed model. To train, validate, and test our model, we prepared the dataset ready for use.

Data Preparation: Rhythm-based ECG records were selected from the PTB-XL dataset. Specifically, AFIB, STACH, SARRH, SBRAD, and PACE were selected because they had 296 and above samples in the dataset. The ones that were not selected had a very small number of samples that would create a significant imbalance in the dataset.

Pre-processing of Data: Data cleansing was carried out to filter and select rhythm-based data relevant to the specific needs. The dataset consisted of two sampling frequencies, 100 Hz and 500 Hz, with each record containing either 1000 or 5000 discrete values depending on the frequency. We chose the 100 Hz samples as it conforms to standards worldwide. We also have normalized the data.

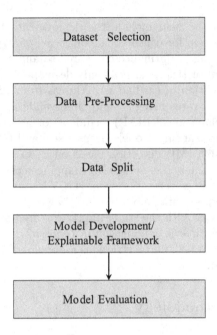

Fig. 3. Methodology Block diagram.

Dataset Splitting: To prepare the data for training, validation, and testing purposes, we applied a technique called "10-fold data splitting." This methodology involves dividing the available dataset into ten folds, with each fold being of equal size. In our specific case, we had approximately 4030 ECG signal records, which were divided into ten folders, each containing approximately 403 records. The assignment of records to the folds was done randomly during runtime, ensuring a fair distribution of data across the folds.

3.2   Model Development

To create an explainable framework for the ECG classification task, we used 12 Resnet-18 models with some modifications, as shown in Fig. 4. The first modification is to make the ResNet-18 work with 1D ECG signals. The second modification collects the features generated by all the 12 models and aggregates them to be passed to one fully connected layer for classification.

The model takes an input of shape (1000, 12), where 1000 represents samples of a continuous signal from a sampling frequency of 100 Hz and a sampling time of 10 s, and 12 represents the number of channels.

We employed the Grad-CAM++ algorithm as an explainable framework to produce heatmaps for highlighting the critical ECG signal regions for our model's classification. The algorithm comprises the subsequent steps: Feature mappings are extracted from the final convolutional layer from each Resnet-18 model, the

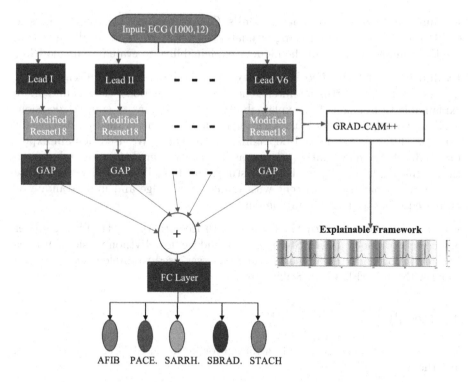

Fig. 4. Diagram of our proposed method.

predicted class's gradient about the feature maps is calculated using backpropagation, the weights of each feature map are determined by averaging the gradient values, the weighted sum of the feature maps is calculated to create a heatmap, the heatmap is activated using a ReLU activation function, and the heatmap is normalized to make it easier to visualize.

Experimental Hyperparameter Setup: A good experimental hyperparameter setup for a DL model is important to avoid under or overfitting problems and find the best results for the given problem. Key parameters are required to optimize the performance of a DL model; these are activation function, learning rate, optimizer, dropout, loss function, number of epochs, and batch size. For the DL model to perform at its best, the hyperparameter variables that control the training process must be adjusted. To determine the best combination of hyperparameters that reduces loss and error, we used iterative trial and error until we found the hyperparameter set at which we got the best model performance. These are RELU for activation function, 0.001 learning rate, Adam optimizer, 0.2 dropouts, categorical-crossentropy loss function, 60 epochs, and a batch size of 32.

Evaluation Metrics for the Model's Classification: The effectiveness of our DL model is measured using evaluation metrics. We used recall, precision, the F1-score, accuracy, and the model's explainability to evaluate our model.

Evaluation of Model Explainability: Explainability is a new discipline that focuses on making artificial intelligence model predictions clearer and more explainable to humans. The explainability technique aims to support decision-making and helps to build trust with the model. Different evaluation metrics for explainability may be needed depending on the goals. We used domain expertise feedback as an evaluation metric, as it was considered the most relevant to the specific explainability goals in our study. Our model's capability to generate explanations that are consistent with human knowledge and understanding is evaluated using feedback from domain experts.

Experimentation Setup: MacBook Pro, Processor: Apple M1, Total Number of Cores: 8; Memory: 16GB; Storage: 1TB; and we used Python version 3.9.12 as the main programming language, and Keras was used to implement the model along with TensorFlow and Scikit-learn.

4   Results

In this part, we present the outs achieved for both the ECG classification task and the explainable framework.

4.1   Results of ECG Classification

Table 1. Classification Report for the five class labels.

Classes	Precision	Recall	F1-score	Accuracy	support
AFIB	0.94	0.92	0.93	0.95	152
PACE	0.95	0.68	0.79	0.98	28
SARRH	0.78	0.90	0.84	0.93	77
SBRAD	0.88	0.92	0.90	0.97	63
STACH	0.96	0.93	0.94	0.98	82
Average	0.90	0.87	0.88	0.96	402

The proposed approach demonstrated a good recall rate of 0.87, indicating a reduced number of false negatives. This suggests that the method effectively identified a significant portion of relevant instances (Table 1). Furthermore, the F1-score, which gives a comprehensive evaluation of the model's performance, yielded a reliable value of 0.88. This F1 score reflects the method's strong balance between precision and recall, emphasizing its overall effectiveness.

4.2   Model Explainability

DL models often lack explainability in real-world applications. To address this, we employed Grad-Cam and Grad-Cam++ techniques to explain our model's predictions.

Visual Explanations for ECG Classification Using Grad-Cam and Grad-Cam++. Here we considered small sample outputs just to compare the two visual explainability measures, Grad-Cam and Grad-Cam++. Both Grad-CAM and Grad-CAM++ are methods for producing visual explanations that highlight the areas of an ECG signal that are crucial for a classification decision made by a deep neural network.

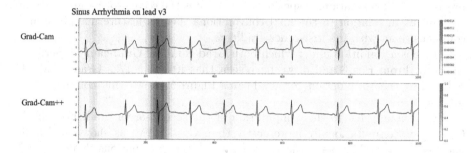

Fig. 5. Grad-cam vs. Grad-cam++ visual explanations for SARRH

Grad-CAM creates a visual explanation by calculating the gradients of a convolutional neural network's output in relation to the last convolutional layer's feature mappings These gradients are then averaged, and the resulting map is utilized to weight the last convolutional layer's feature mappings. To create the final visual explanation, the weighted feature maps are then added together.

In addition to the first-order gradients present in Grad-CAM, Grad-CAM++ expands Grad-CAM by integrating second-order gradients. To produce a more

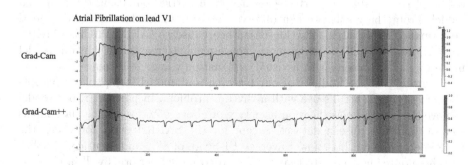

Fig. 6. Grad-cam vs. Grad-cam++ visual explanations for AFIB on lead V1

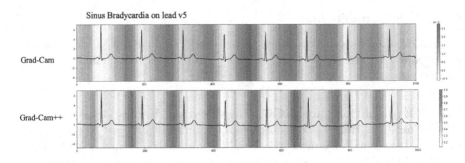

Fig. 7. Grad-cam vs. Grad-cam++ visual explanations for SBRAD on lead V5

precise visual explanation, it specifically computes the second-order gradients of the output in relation to the final convolutional layer's feature mappings and combines them with the first-order gradients.

There is a condition under which both Grad-cam and Grad-cam++ perform similarly. From Eqs. 3 and 6, we have seen that if $\alpha_{ij}^{kc} = 1/Z$, Grad-CAM++ reduces to the formulation for grad-cam. Figure 5 shows this condition has occurred.

Even though due to the computation of the second-order gradients, Grad-Cam++ is computationally more costly than Grad-CAM, as shown in Fig. 6, and Fig. 7, Grad-CAM++ typically generates visual explanations that are more precise and concentrated than Grad-CAM; that is, Grad-Cam++ is better than Grad-Cam in visual explanations and important region localization.

From Fig. 8 a normal sinus rhythm has consistent and regularly spaced wave patterns and R-R intervals. Sinus arrhythmia (SARRH): on Lead III and Lead V6, there is a variation to the normal sinus rhythm that is characterized by an irregular rate and changing R-R intervals. Our model used deep colors to show the most significant region in the ECG signals used for the classification, with a considerable change in the R-R intervals.

Expert Evaluation of the Model Explanations. In this work, a classification of rhythm-based heart disease has been carried out using the proposed model. From the result, we can observe that our model has achieved a good performance in classifying rhythm-based heart disease (AFIB, PACE, SARRH, ABRAD, and ATACH). Furthermore, we were able to reason out why our model made the classifications with a visual explanation using Grad-cam++.

According to the findings of our research, our model is capable of acceptably classifying heart disease with a rhythm problem. In particular, our model achieved an overall classification accuracy of above 95% for all the class labels contained in the unseen dataset, except for the SARRH class, for which the model determined a little lower average F1-Score of 88% and accuracy of 93%.

These outcomes indicate that the proposed model performs well in classifying and distinguishing between different types of cardiac diseases based on their

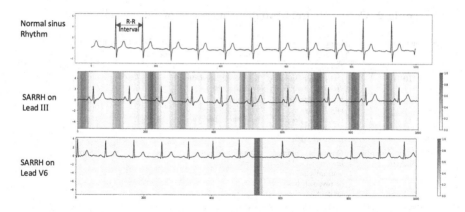

Fig. 8. Normal sinus Rhythm and sinus arrhythmia on leads II and v6

characteristic rhythms. Our model's accuracy and the explanation for its classification suggest that it can be used as a reliable tool to support the identification of people with these diseases.

Overall, our results provide important insights into the effectiveness of deep learning approaches, such as the Resnet-18 model, in the field of healthcare when the explainability framework is included in the workflow of deep learning.

The standard visual explanation methods we used include Grad-cam and Grad-cam++. It was found that Grad-cam++ produced better visual explanations and more precise localization of important regions when results from using Grad-cam and Grad-cam++ for visual explanations were compared. In other words, Grad-cam++ outperformed Grad-cam in its ability to pinpoint the most crucial areas of the input data that influenced the model's decision-making.

Based on the results from the two domain experts given in Table 2, it can be said that the output of the model is mostly correct and consistent with the interpretations of the two domain experts. The fact that both doctors had issues with the classification of the PACE record reflects the limited number of records used for training and testing in this class. Their issues also agreed with the smallness values of 0.68 on the recall scale and 0.79 on the F1-score found in the result for PACE, which is the smallest when compared to other classes' recall values and F1-scores.

Overall, the use of visual explanations such as Grad-cam and Grad-cam++ can produce valuable understandings of the decision-making process of deep learning models and help improve their transparency and interpretability.

Limitations: In this study, only five heart diseases related to rhythm problems were identified by our model. Other classes of rhythm-based heart diseases were not addressed. Furthermore, the results of our model were not fully validated by specialists, and only ten samples of the results were evaluated by two domain experts.

Table 2. Comments from two domain experts on our model's results

Records	Condition	Expert's Comment		Agree, Disagree or Equivocal	
		Expert One	Expert Two	Expert One	Expert Two
1	Sinus arrhythmia (SARRH)	Comments: Sinus arrhythmia, baseline artifacts, poor R wave progression or late R wave transition, possibly counterclockwise heart rotation	Sinus arrhythmia: Equivocal: No strong evidence; the threshold for irregularity needs to be a bit higher	Agree	Equivocal
2	Sinus bradycardia (SBRAD)	Comments: Sinus bradycardia, likely lead misplacement or reversal, baseline artifact	Sinus bradycardia: Agree	Agree	Agree
3	Atrial fibrillation (AFIB)	Comments: Atrial fibrillation with fast ventricular response	Atrial fibrillation: Agree	Agree	Agree
4	Normal functioning artificial pacemaker (PACE)	Comments: Atrial fibrillation; impossible to see pacemaker spikes (if present) because of artifacts	Normal functioning artificial pacemaker (PACE): Equivocal (no visible clue for pacing but could be a paced rhythm like in bipolar VVI pacemakers): Irregularity due to premature ventricular contraction	Disagree	Equivocal
5	Sinus bradycardia (SBRAD)	Comments: Sinus bradycardia, baseline artifact, benign early repolarization	Sinus bradycardia (SBRAD): Agree	Agree	Agree
6	Sinus tachycardia (STACH)	Comments: Sinus tachycardia, PVCs (premature ventricular complexes)	Sinus tachycardia (STACH): Agree (with single PVC on second QRS)	Agree	Agree
7	Atrial fibrillation (AFIB)	Comments: Multifocal atrial fibrillation	Atrial fibrillation: Agree (with fast ventricular response)	Agree	Agree
8	Sinus arrhythmia (SARRH)	Comments: baseline artifacts and baseline wander. Sinus Arrhythmia	Sinus arrhythmia (SARRH): Agreed, but might still be normal if slight irregularities are ignored (same as above)	Agree	Agree
9	Sinus tachycardia (STACH)	Comments: Sinus tachycardia, aberrantly conducted PAC	Sinus tachycardia: Agree (with PVC)	Agree	Agree
10	Atrial fibrillation (AFIB)	Comments: Atrial fibrillation	Atrial fibrillation: Agree	Agree	Agree

5   Conclusion

Our work aimed to develop an explainable model for classifying rhythm-based heart disease from 12-lead ECG signals using the PTB-XL dataset. We proposed an approach using the ResNet-18 deep learning model integrated with visual explainers (Grad CAM and Grad CAM++). We examined the impact of employing Grad-Cam and Grad-Cam++ visual explainers on the reliability of classifying rhythm-based cardiac disease. The findings demonstrate that our technique effectively categorizes rhythm-based heart disease, while the visual explainability builds trust in the deep learning model's predictions. Our model was effective, with a classification accuracy of 96% and an F1 score of 0.88. We gave samples of our model's explanations to two domain experts and the experts agreed with 80% of the explanations given to them. In general, our work addressed the need for an explainable model in cardiac diagnostics, and valuable insights were gained about visual explainers.

6 Future Work

In future work, it is recommended to include all remaining rhythm types in the study, as the current research focused on only five out of eleven rhythm-based heart diseases due to limited data availability. Gathering a larger range of ECG recordings from different datasets would be beneficial for expanding the training dataset. This expansion has the potential to improve the explanation of the classification of all rhythm-based heart problems if the remaining six rhythms become available.

Collaboration with domain experts is important to ensuring the practical application of research results in the real world. Their expertise and experience can contribute to the successful implementation of the findings if they validate all the classifications made by the deep learning model.

References

1. Ayano, Y.M., Schwenker, F., Dufera, B.D., Debelee, T.G.: Interpretable machine learning techniques in ECG-based heart disease classification: a Systematic Review. Diagn. **13**(1), 111 (2022). https://doi.org/10.3390/diagnostics13010111
2. Chattopadhay, A., Sarkar, A., Howlader, P., Balasubramanian, V.N.: Grad-cam++: generalized gradient-based visual explanations for deep convolutional networks. In: 2018 IEEE Winter Conference on Applications of Computer Vision (WACV), pp. 839–847. IEEE (2018)
3. Cheffer, A., Savi, M.A., Pereira, T.L., de Paula, A.S.: Heart rhythm analysis using a nonlinear dynamics perspective. Appl. Math. Model. **96**, 152–176 (2021)
4. Choi, E., Bahadori, M.T., Sun, J., Kulas, J., Schuetz, A., Stewart, W.: Retain: an interpretable predictive model for healthcare using reverse time attention mechanism. In: Advances in Neural Information Processing Systems **29** (2016)
5. Choi, K., Fazekas, G., Sandler, M., Cho, K.: Convolutional recurrent neural networks for music classification. In: 2017 IEEE International Conference on Acoustics, Speech and Signal Processing (ICASSP), pp. 2392–2396. IEEE (2017)
6. Feyisa, D.W., Debelee, T.G., Ayano, Y.M., Kebede, S.R., Assore, T.F.: Lightweight multireceptive field CNN for 12-lead ECG signal classification. Comput. Intell. Neurosci. **2022**, 1–14 (2022). https://doi.org/10.1155/2022/8413294

7. Ganeshkumar, M., Ravi, V., Sowmya, V., Gopalakrishnan, E., Soman, K.: Explainable deep learning-based approach for multilabel classification of electrocardiogram. In: IEEE Transactions on Engineering Management (2021)
8. Hannun, A.Y., et al.: Cardiologist-level arrhythmia detection and classification in ambulatory electrocardiograms using a deep neural network. Nat. Med. **25**(1), 65–69 (2019)
9. Hong, S., Zhou, Y., Shang, J., Xiao, C., Sun, J.: Opportunities and challenges of deep learning methods for electrocardiogram data: a systematic review. Comput. Biol. Med. **122**, 103801 (2020)
10. Izci, E., Ozdemir, M.A., Degirmenci, M., Akan, A.: Cardiac arrhythmia detection from 2D ECG images by using deep learning technique. In: 2019 Medical Technologies Congress (TIPTEKNO), pp. 1–4. IEEE (2019)
11. Jahmunah, V., Ng, E.Y.K., Tan, R.S., Oh, S.L., Acharya, U.R.: Explainable detection of myocardial infarction using deep learning models with grad-cam technique on ECG signals. Comput. Biol. Med. **146**, 105550 (2022)
12. Jing, E., Zhang, H., Li, Z., Liu, Y., Ji, Z., Ganchev, I.: ECG heartbeat classification based on an improved resnet-18 model. Comput. Math. Methods Med. **2021** (2021)
13. Kumar, I., Scheidegger, C., Venkatasubramanian, S., Friedler, S.: Shapley residuals: quantifying the limits of the shapley value for explanations. Adv. Neural. Inf. Process. Syst. **34**, 26598–26608 (2021)
14. Muhammad, M.B., Yeasin, M.: Eigen-CAM: class activation map using principal components. In: 2020 International Joint Conference on Neural Networks (IJCNN), pp. 1–7. IEEE (2020)
15. Selvaraju, R.R., Cogswell, M., Das, A., Vedantam, R., Parikh, D., Batra, D.: Grad-CAM: Visual explanations from deep networks via gradient-based localization. In: Proceedings of the IEEE International Conference on Vvision, pp. 618–626 (2017)
16. Smulyan, H.: The computerized ECG: friend and foe. Am. J. Med. **132**(2), 153–160 (2019)
17. Soliński, M., et al.: 12-lead ECG arrythmia classification using convolutional neural network for mutually non-exclusive classes. In: 2020 Computing in Cardiology, pp. 1–4. IEEE (2020)
18. Somani, S., et al.: Deep learning and the electrocardiogram: review of the current state-of-the-art. EP Europace **23**(8), 1179–1191 (2021)
19. Wagner, P., et al.: PTB-XL, a large publicly available electrocardiography dataset. Sci. data **7**(1), 1–15 (2020)
20. Yin, W., Kann, K., Yu, M., Schütze, H.: Comparative study of CNN and RNN for natural language processing. arXiv preprint arXiv:1702.01923 (2017)
21. Zhang, D., Yang, S., Yuan, X., Zhang, P.: Interpretable deep learning for automatic diagnosis of 12-lead electrocardiogram. Iscience **24**(4), 102373 (2021)

Development of an Explainable Heart Failure Patients Survival Status Prediction Model Using Machine Learning Algorithms

Betimihirt Getnet Tsehay Demis(ORCID) and Abdulkerim M. Yibre[✉](ORCID)

Department of Information Technology, Faculty of Computing, Bahir Dar Institute of Technolgy, Bahir Dar University, Bahir Dar, Ethiopia
abdukerimm@gmail.com

Abstract. Heart failure is a condition where insufficient blood circulation occurs throughout the body, causing the heart to become weak or stiff and not function properly. This can lead to congestive heart failure. The life expectancy of a congestive heart failure patients can vary depending on factors such as disease severity, age, genetics, and other variables. The Centers for Disease Control and Prevention (CDC) has reported that almost half of the patients with congestive heart failure survive for more than five years. Previous studies have attempted to predict the survival rate of heart failure patients, but most of them have been limited by unbalanced data sets and a small number of examples. The main goal of this research is to develop a machine learning model that can accurately predict the survival status of heart failure patients while providing an insight into its decision-making process. Machine learning algorithms are a part of artificial intelligence systems and can be used to predict output values based on input data. However, Many machine learning models lack an explanation of how they arrive at their decisions. Therefore, explainable machine learning algorithms are now being used to understand and explain the decision-making process of machine learning models. In this research, we not only use the latest machine learning algorithms for prediction, but we also incorporate explainability features to understand the results. Hyper-parameter optimization techniques are employed to find the best parameter values for better predictive accuracy. Our data set consists of 1334 heart failure patients' data collected from Felege Hiwot referral hospital and Injibara general hospital. The prediction models used include Decision Tree, Linear Regression, K-Nearest Neighbor, Deep Neural Network, and XGBoost. Prior to conducting the actual experiment, the data set is balanced using the Synthetic Minority Oversampling Technique (SMOTE). Experimental findings indicate that XGBoost performs better than other machine learning techniques, achieving an AUC value of 0.93 using Grid Search HPO after SMOTE.

Keywords: Heart Failure · Survival Event · Follow-up Period · Synthetic Minority Oversampling Technique · Hyper-parameter Optimization · Model Explainability

© The Author(s), under exclusive license to Springer Nature Switzerland AG 2024
T. G. Debelee et al. (Eds.): PanAfriConAI 2023, CCIS 2068, pp. 117–137, 2024.
https://doi.org/10.1007/978-3-031-57624-9_7

1   Introduction

All organs must function properly for a person's life to continue. The heart is the most crucial organ in the human body and is responsible for the vital function of pumping and circulating the blood, which carries the necessary oxygen and nutrients needed by the body. However, the heart is affected by several diseases, collectively known as cardiovascular diseases. Cardiovascular diseases (CVDs) refer to a collection of conditions that affect the blood vessels and heart. It comprises a variety of illnesses, such as cardiac attack, cerebrovascular accident, heart failure, and hypertensive and rheumatic heart ailments. According to the WHO (World Health Organization), CVDs are the leading cause of death globally [11]. In 2019, an approximate of 17.9 million deaths were attributed to CVDs, accounting for 32% of all global fatalities, with heart attacks and strokes being accountable for 85% of these fatalities. Most deaths due to CVDs happen in countries that have low to middle economic status [11].

Heart failure happens when the heart is not capable of sufficiently circulating blood throughout the body. This condition is identified by symptoms such as difficulty breathing, consistent coughing or wheezing, swollen ankles, and fatigue. It is responsible for a significant number of hospital visits and a decreased quality of life. To predict survival in patients with heart failure, the most relevant indicators are serum creatinine and ejection fraction [8]. Heart failure is a rapidly expanding public health concern and a leading cause of death worldwide. It is a pandemic that is spreading across the globe and is affecting approximately 40 million people worldwide [23]. A hospital-based mortality study in Ethiopia that included a sample of all age groups found that HF was responsible for 2.5% of fatalities [2].

Survival status of heart failure patients refers to whether a patient is alive or deceased before the end of their follow-up period. In other words, it determines the outcome of the patient's condition during receiving treatment. A heart failure patient who survives is still alive and receiving medical care, while a heart failure patient who dies has passed away as a result of the heart failure condition. Prediction of survival status can be used to assess the effectiveness of treatment methods and to develop better strategies for managing heart failure patients.

Predicting the probability that a heart failure patient will survive is crucial for reducing the fatality of heart failure patients' by knowing its causes. Therefore, in this research, we have employed machine learning algorithms to forecast the patients' survival status concerning heart failure. The data is collected from Felege Hiwot referral hospital and Injibara general hospital in Ethiopia.

The primary objective of this study is to develop an explainable model utilizing machine learning techniques that could support healthcare professionals in hospitals in predict the survival status of heart failure patients before the end of the follow-up period. Moreover, the model can help them to comprehend the characteristics that encourage patient survival or mortality when rendering care.

This research predicts the survival status of individuals suffering from heart failure, which helps to overcome the problem of healthy people in society by predicting their survival event. It allows the Ministry of Health in Ethiopia to receive scientific information about the findings, which assists policymakers in raising societal awareness about the survival event of heart failure patients. The result of this research also efficiently and successfully (effectively) helps doctors predict the survival events of heart failure patients before the end of the follow-up period. To better treat patients before they die, doctors use this prediction model to determine how many heart failure patients have passed away or survived before the end of their follow-up period.

This study provides significant contributions for academic researchers exploring the prediction of heart failure patient survival by establishing the fundamental characteristics needed for such prognostications. The features are listed in descending order based on their mutual information score and by preparing a data set. We have collected a much larger data set than the one used in similar studies earlier. Another contribution of this research is that we have employed SMOTE to balance the data set and utilize various hyper-parameter optimization techniques (including Bayesian, Grid search, and Random search) to select the optimal hyper-parameter combinations and enhance the classification model's performance. We have also done model explainability for the XGBoost model using SHAP and LIME techniques. This explains the contribution of each feature to the prediction of each class (Survive and Death) and defines how the model works on prediction tasks.

The remaining part of this paper includes a problem statement in Sect. 2, Related Works are included in Sect. 3, Materials and Methods in Sect. 4, Results and Discussion in Sect. 5 and Conclusion in Sect. 6.

2   Problem Statement

Heart failure develops when the body's demands cannot be satisfied because the heart is unable to pump enough blood [9]. Heart failure is a significant global public health issue and the primary reason for mortality globally. It affects a lot of individuals [14]. Since so many people suffer from heart failure, it is crucial to predict the survival status of heart failure patients by analyzing their data. Heart failure affects a lot of people in Ethiopia and often results in fatalities. In Ethiopia HF was responsible for 2.5% of fatalities [2]. Therefore, predicting the probability that survival of heart failure is crucial for reducing this fatality by knowing its causes. Predicting the survival status of individuals with heart failure helps to assess the effectiveness of medications and treatments. Through the assessment of the effectiveness of treatments, Healthcare professionals can more effectively personalize care treatments for individual heart failure patients and achieve better clinical results. A treatment effectiveness assessment might point out areas that require additional study and the development of new medicines.

Without the aid of advanced computational technology, it might be challenging to predict the survival status of patients with heart failure. Therefore, a model that addresses this issue must exist. Hence, in this study, we developed an explainable model that can learn from HF patients' clinical data and predict the survival status of patients or whether the patient died or survived before the end of the follow-up period and that can identify attributes that contribute to patients survival or death to help doctors more effectively and efficiently using machine learning techniques. The majority of them have used 299 size of data and did not consider balancing the data set [1,3,8,9,19]. To improve the model's performance and its generalizability, we increased our data set, utilizing an imbalance data set handling technique, and applying hyper-parameter optimization methods. We have also used explainability techniques to gain insight into the process by which a model is producing its predictions. The prediction of diseases is significantly aided by machine learning algorithms. Machine learning can predict future data or draw a more desirable conclusion from a data set.

3   Related Works

There is a lot of research done on predicting the survival events of heart failure patients using machine learning, and such works produce outstanding outcomes. Many Researchers have published their work over the years, and each of them has made a significant contribution to the prediction of the survival status of heart failure patients. There is research dedicated to enhancing the accuracy of predicting heart failure patients' survival rates, involving identifying the optimal machine learning approach [9], while others set their contribution by identifying factors influencing the survival time of heart failure [1,12,23]. The following current studies on heart failure patient survival event prediction can be useful sources of information for this research. Moreno-Sanchez [16] has used RF, DT, Extra Trees, Gradient Boosting, AdaBoost, and XGBoost algorithms to develop an explainable prediction model of heart failure survival using ensemble trees. He used 299 samples of data set and they didn't balance their data set. Ishaq et al. [13]have done their research on improving the prediction of heart failure patients' survival using SMOTE and effective data mining techniques, and their study employs nine classification models: Gaussian Naive Bayes classifier (G-NB), Stochastic Gradient classifier (SGD), Gradient Boosting classifier (GBM), DT and Support Vector Machine (SVM). Their experiment result achieved 82% of accuracy with DT, 54% of accuracy with SGD, 74% of accuracy with G-NB, and 76% of accuracy with SVM.

Chicco & Jurman [8] have used a data set of 299 patients with heart failure collected in 2015. They apply several machine learning classifiers such as RF, Linear Regression, ANN, SVM, DT, k-Nearest Neighbors, Naïve Bayes, and Gradient Boosting to predict the patients' survival and rank the features corresponding to the most important risk factors. Their findings indicated that the most significant characteristics were serum creatinine and ejection fraction, based on which they constructed machine learning models to predict the survival rate of individuals with congestive heart failure.

Almazroi [1] The Primary objective was to identify crucial attributes that could accurately forecast the likelihood of developing heart-related illnesses and estimate the chances of survival. Almazroi [1]used 299 patients' data and he split 80% for training and 20% for testing, and SVM, LR, ANN, and DT algorithms.

According to his study, the decision tree algorithm (with a success rate of 80%) outperformed logistic regression (with a success rate of 78.34%), support vector machines (with a success rate of 66.67%), and artificial neural networks (with a success rate of 60%). This means that decision trees had a 14% higher accuracy than the combined average accuracy of logistic regression, support vector machines, and artificial neural networks. Nonetheless, it's important to note that the study did not consider balancing the data set. Foziljonova & Wasito [9] have identified the most suitable approach based on machine learning to forecast the rate of survival in patients with heart failure. They used 299 patients' data and Random Forest (RF) and K – nearest Neighbors (KNN). Their result shows that they achieved 82% accuracy with KNN.

4   Materials and Methods

4.1   The Proposed Prediction Model

Figure 1 Shows the general approach of the heart failure patients' survival status prediction model. This model includes seven major parts. The first is data collection from Injibara General Hospital and Felege Hiwot Referral Hospital. The second part is data pre-processing, which improves our data quality and includes data transformation, filling missing values, feature selection, and data balancing tasks. The third part is data set splitting, which splits our data set into training, validation, and testing sets. The fourth part is model building and training using selected algorithms (DT, LR, KNN, DNN, and XGBoost). The fifth part is model evaluation metrics, which evaluate the model's performance, the sixth part is model selection, and the last part is applying model explainability for selected models.

4.2   Data Collection

In this study, we have collected data from Felege Hiwot Referral Hospital and Injibara General Hospital. We used documents as data collection sources to gather relevant data for the study. Different studies have used age, gender, smoking, diabetes, anemia, blood pressure, serum sodium, serum creatinine, platelets, follow-up time, ejection fraction, Creatine phosphokinase (CPK), and event attributes to predict the survival status of heart failure patients, and set, physical environment, and blood type, which include information about patients' physical characteristics, can help in recognizing additional contributors to the risk of cardiovascular diseases. Based on this and with the help of professional guidance, we have collected data with 14 attributes, which are age, gender, smoking, diabetes, anemia, blood pressure, serum sodium, serum creatinine, platelets, physical environment, blood type, follow-up time, ejection fraction, and status. Because we can't find Creatine phosphokinase (CPK) on patients' cards, it is not part of our data set.

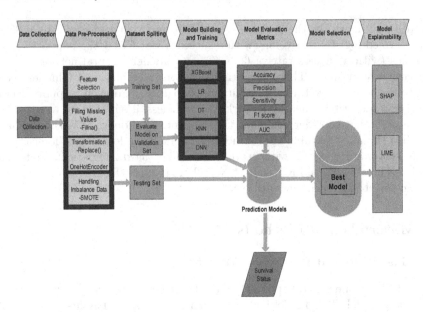

Fig. 1. Heart Failure Patients Survival Status Prediction Model

Data Description. Table 1 shows the attributes that are included in our data set along with their descriptions and data types.

4.3   Data Pre-processing

It is vital to undertake pre-processing operations because raw data are susceptible to noise, corruption, missing data, and inconsistent data. Thus, the best approach to solving these issues is data pre-processing. It is the most crucial and significant aspect of a supervised machine learning algorithm's generalization performance [15]. Missing values, noise, inconsistencies, and redundancy are anomalies that are typically present in raw data and have an impact on how well subsequent learning processes work. Thus, a suitable pre-processing step is frequently carried out to reduce the impact of data abnormalities on the effectiveness of subsequently processed steps [18].

The data must first go through pre-processing to create the correct format and data type in order to predict the survival status of heart failure patients. For the classification process to work more accurately and efficiently, data preparation is essential. As a result, data transformation, feature selection, and handling of class imbalance in the data set has been conducted.

Table 1. Collected attributes with their description

No	Attribute Name	Description	Data Types
1	Age	Age of the patient in years	Numeric
2	Gender	The patient is a woman or man	Categorical
3	Smoking	Whether the patient smokes or not	Categorical
4	Diabetes	Does a patient have diabetes?	Categorical
5	Blood pressure	Does a patient have hypertension?	Categorical
6	Anemia	Does a patient have decreased level of red blood cells?	Categorical
7	Ejection Fraction	The percentage of blood that leaves the heart with each contraction	Numeric
8	Serum Sodium	Level of sodium in the blood	Numeric
9	Serum Creatinine	Level of creatinine in the blood	Numeric
10	Platelets	Level of platelets in the blood	Numeric
11	Physical Environment	Patients' Physical Environment	Categorical
12	Blood Type	Patients' Blood Type	Categorical
13	Follow-Up Time	Patient follow-up time in days	Numeric
14	Status	Does a patient die or survive during the follow-up period	Categorical

Data Transformation

Our data set contains 14 attributes, with the class attribute 'Status' being one of them. Certain attributes like Gender, Smoking, Diabetes, Blood Pressure, Anemia, Physical Environment, Blood Type, and Status have categorical values and Age, Ejection Fraction, Serum Sodium, Serum Creatinine, Platelets and Follow-Up Time have Numeric values. To enable experimentation, we transformed these categorical values into numerical ones. The Gender and Physical Environment attributes were converted using OneHotEncoder() method, whereas the remaining attributes were converted using the replace() method. The target class, which is the status attribute, was assigned 1 for 'Survive' and 0 for 'Death' Smoking, Diabetes, Blood Pressure, and Anemia attributes were assigned 1 for 'Yes' and 0 for 'No'. Blood Type, which had multiple values (A, A+, A−, B, B+, B−, AB, AB+, AB−, O, O+, O), was replaced with numerical values (1, 2, 3, 4, 5, 6, 7, 8, 9, 10, 11, and 12) respectively (Fig. 2).

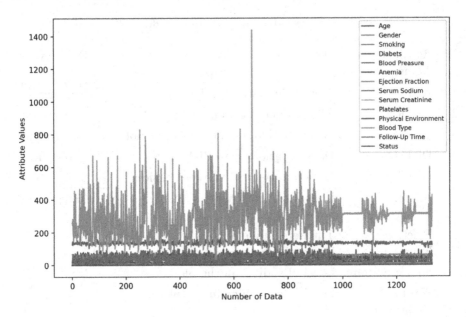

Fig. 2. Attributes with their values

Feature Selection

Finding the best representative subset of characteristics for the prediction task can help increasing prediction accuracy, decrease computation time, and reduce over-fitting and model complexity [7]. In this research, we have selected the best features that have strong dependence on the target variable Status by using mutual_info_classif() function. This function is used to calculate the mutual information score between each feature and the target (class) attribute. Mutual information is a measure of the statistical dependence between two variables, each feature and class variable. A higher mutual information score indicates a higher level of information gain between the feature and the target variable, and therefore a stronger relationship between them. Based on this we have selected 10 features that have a strong dependence with the target variable which are Follow-Up Time, Physical Environment, Serum Creatinine, Smoking, Diabetes, Serum Sodium, Age, Platelets, Ejection Fraction, and Blood Pressure. In Table 2 the mutual information scores of each feature with the class attribute (Status) are displayed in descending order from highest to lowest.

Data Set Splitting

The most crucial step in developing a supervised model is model validation. A valid data-splitting method is necessary for model validation and the construction of a model with good generalization performance. The data must be divided into training, validation, and testing sets to avoid over-fitting. The validation set

Table 2. Mutual information score for each Feature Sorted in descending order

Rank	Feature − Names	mutual inf oscores
12	Follow-Up Time	0.047740
10	Physical Environment	0.032419
8	Serum Creatinine	0.025659
2	Smoking	0.015305
3	Diabetes	0.014220
7	Serum Sodium	0.013222
0	Age	0.011448
9	Platelets	0.006471
6	Ejection Fraction	0.006465
4	Blood Pressure	0.004991
1	Gender	0.000000
5	Anemia	0.000000
11	Blood Type	0.000000

is used to validate each trained model after it has been constructed using a variety of model parameter settings on the training set. In the split between training, validation, and testing, the model is trained to learn any hidden patterns in the data. The validation set is a set of data that is separate from the training set and is used to verify the performance of our model as it is being trained. The primary aim of separating the data set into a validation set is to avoid overfitting the model, which occurs when the model becomes highly skilled in classifying the examples in the training set. Upon the conclusion of the training process, the model undergoes testing on the test set. In our study, we distributed the data set into three parts: 60.

Handling of Class Imbalance Data Set

In class imbalance, the majority of the cases belong to one class, while much fewer examples belong to the other class, which is typically the more significant class. In this scenario, standard machine learning algorithms often struggle to handle the underrepresented class as they tend to prioritize achieving precise results for the majority class. As a result, typical classifiers may overlook the minority class [10]. In this study, we used the over-sampling mechanism, SMOTE, to address class imbalance. In SMOTE, instead of simply copying existing ones, new artificial instances from the minority class are created [6]. SMOTE overcomes the over-fitting issue and causes the decision boundaries for the minority class to extend wider into the majority class space by using interpolation rather than replication [10]. As indicated in [25], after balancing, the total number of instances (T) exceeds the size of the imbalanced data set as given in Eqs. (1) and (2).

$$T = m_c + n_c + n'$$ (1)

$$n'_c = (r_c - 1)n_c$$ (2)

4.4 Hyper-parameter Optimization

Hyper-parameter optimization (HPO) is the process of finding the most favor-able combinations of hyper-parameters that can produce the best performance on the given data. In order for any machine learning algorithm to achieve opti-mal functionality, it is necessary to adjust its hyper-parameters [20]. Different types of hyper-parameters are included in advanced machine learning algorithms, including XGBoost, RF, DT, SVM, and deep neural networks (DNN), and their tuning directly affects the algorithm's performance [24].

The hyper-parameter default settings are provided with the majority of machine learning algorithms. However, on different kinds of machine learning projects, the default values don't always work effectively. Hence, the hyper-parameters need to be optimized in order to get the right combination that will provide the best performance. Hence, in this study, we have used Bayesian opti-mization, grid search, and random search hyper-parameter optimization meth-ods.

Bayesian optimization iteratively runs models with various sets of hyper-parameter values while analyzing the information from past models to choose hyper-parameters for the newest model. The algorithm becomes increasingly confident in determining which areas of parameter space are worth examining and which are not as the number of observations increases [5]. Grid search is a common method for hyper-parameter tuning that works by trying every possible combination of the parameters to find the best values. In grid search, a subset of the hyper-parameters of a machine learning algorithm are manually defined and then searched. With the addition of more hyper-parameters, grid search proves to be computationally expensive [21]. In the random search approach, the opti-mal solution for the created model is found by using random combinations of the hyper-parameter values. A generative method is used in random search to create random samples. The configuration space is defined using this genera-tive method, and the assignments for the hyper-parameters are subsequently drawn [4].

4.5 Model Explainability

Model explainability refers to techniques, methods, and tools used to understand the inner workings of a machine learning model and the reasoning behind its predictions. The main purpose of model explainability is to build trust in the predictive power of the model and to ensure that its decisions are understandable. Explainable machine learning models empower healthcare professionals to make informed decisions based on data analysis. By using such models, personalized treatment can be provided, which ultimately results in better healthcare service quality [22].

4.6   Model Evaluation

It is responsible for defining the evaluation criteria and results for the specified model. We have used a confusion matrix and performance measures such as accuracy, precision, sensitivity (recall), F1 score, and ROC-AUC (Receiver Operating Characteristic-Area under Curve) to evaluate the performance of our model.

Confusion Matrix

One of the most direct and basic metrics for assessing the accuracy and correctness of the model is the confusion matrix.

- True Positive (TP) refers to the count of instances that are both predicted and actual outcomes of survival.
- True Negative (TN): refers to the count of instances where both the predicted outcome and the actual outcome are that of death
- False Positive (FP): refers to the count of instances where the actual outcome is death but incorrectly predicted as survival
- False Negative (FN): refers to the count of situations where the actual outcome is survival but the predicted outcome is death incorrectly.

Performance Measure

Evaluation metrics measure how well a predictive model performs. In classification tasks, accuracy is the metric that is most frequently used. However, the accuracy statistic can be misleading in the case of unbalanced data sets [17].

- Accuracy: The sum of the accurately predicted "survive" cases and "death" cases, divided by the total number of cases.

$$Accuracy = \frac{TP + TN}{TP + FP + TN + FN} \qquad (3)$$

- Precision: The number of accurately predicted "survive" cases divided by the sum of accurately predicted "survive" cases and inaccurately predicted "survive" cases.

$$Precision = \frac{TP}{TP + FP} \qquad (4)$$

– Sensitivity: The number of "survive" cases predicted correctly divided by the sum of accurately predicted "survive" cases and incorrectly predicted "death" cases.

$$Sensitivity = \frac{TP}{TP + FP} \tag{5}$$

– F1 score: The F1 score combines recall and accuracy into one and is defined as the harmonic mean of precision and recall.

$$F1score = 2 * [(precision * recall)/(precision + recall)] \tag{6}$$

– ROC_AUC: ROC is a performance metric commonly used in classification tasks. In the area under the curve (AUC) of the ratio of true positive rates (TPR) to false positive rates (FPR) a greater TPR is preferred, while a lower FPR score is preferable. AUC values can be between 0.5 and 1 More points are preferable [17].

5   Result and Discussion

In this research, we have carried out different experiments using different classification algorithms. We have discussed feature selection, model evaluation metrics before SMOTE, Parameter value spaces and best hyper-parameters after SMOTE, model evaluation metrics with HPO techniques after SMOTE, and model explainability. In addition to accuracy, the models were assessed using different metrics: precision, sensitivity, F1 score, and ROC-AUC, which were computed using the confusion matrix for each algorithm. The original dataset before SMOTE was 1334; we splitted it into 60% for training (1080), 20% (120) for validation, and 20% (134) for testing to build the classification models. To handle class imbalance, we have used SMOTE. After SMOTE the size of dataset was increased to 2034, and we have splitted this balanced dataset into 60% (1301) for training, 20% (326) for validation, and 20% (407) for testing.

5.1   Results of Model Evaluation Metrics Before SMOTE

There is a class imbalance between survive and death classes. So, the models are biased towards survive class and performs poorly on Death class. Table 3 shows the results of Model Evaluation Metrics before SMOTE.

5.2   Parameter Value Spaces and Best Hyper-parameters After SMOTE

During this experiment, we employed several techniques for HPO, including Bayesian, Grid Search, and Random Search, in combination with SMOTE. These techniques were used to identify the parameter value spaces and the best hyper-parameters for each model. Best hyper-parameter combinations for XGBoost model with Grid Search are Colsamplebytree = 0.5, Learningrate = 0.0000001, maxdepth = 9, nestimatores = 50, regalpha = 0.01, reglambda = 0.01, subsample = 1.0 and seed = 0.

Table 3. Results of Model Evaluation Metrics before SMOTE

Model	Accuracy	Precision	Sensitivity	F1_score	AUC
LR	76.1	59	76	66	57
DT	71.3	69	73	71	63
KNN	74.6	67	75	69	69
DNN	74.5	59	7	67	56
XGBoost	79.1	72	77	71	66

5.3   Results of Model Evaluation Metrics After SMOTE with HPO

There is no class imbalance between Survive and Death classes and models are better at identifying both Survive and Death classes. Table 4 shows the results of model evaluation metrics after SMOTE with Bayesian, Grid and Random search HPO techniques. As Fig. 3 and Fig. 4 shows, the XGBoost model achieved a higher AUC value which is 0.93 than other models with Grid Search HPO after SMOTE.

Table 4. Results of Model Evaluation Metrics after SMOTE with HPO

HPO	Model	Accuracy	Precision	Sensitivity	F1_score	AUC
Bayesian search	LR	80.3	81	80	80	84
	DT	83.8	86	84	84	88
	KNN	83.5	88	84	83	88
	DNN	82.8	84	81	81	87
	XGBoost	84.0	88	84	84	89
Grid Search	LR	80.3	81	80	80	84
	DT	84.3	88	84	84	88
	KNN	84.8	88	85	84	89
	DNN	84.2	84	82	82	87
	XGBoost	86.5	88	86	86	93
Random Search	LR	80.1	81	80	80	84
	DT	84.3	83	82	82	87
	KNN	84.8	88	85	84	89
	DNN	82.3	84	81	81	86
	XGBoost	86.5	87	85	85	93

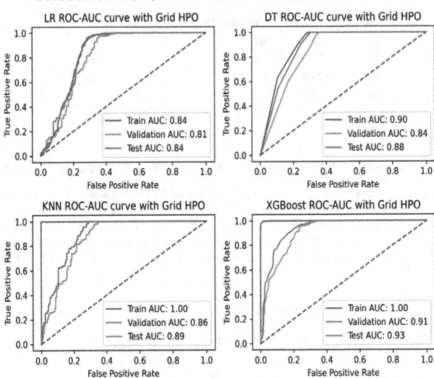

Fig. 3. ROC-AUC curves for LR, DT, KNN and XGBoost algorithms with Grid Search HPO and after SMOTE.

5.4   XGBoost Model Explainability

Using LIME

LIME explains the model's prediction for a given instance by generating a locally faithful, interpretable linear model around that instance. By doing so, LIME highlights which features or attributes of the instance were the most important in resulting in the predicted class probabilities. In Fig. 5 The prediction probability of death class being 0.14 and survival class being 0.86 indicates that in the XGBoost model, the probability of a person surviving is 0.86 (or 86%), while the probability of them not surviving or dying is 0.14 (or 14%). The prediction probabilities represent the estimated likelihood or confidence that the given instance belongs to each possible class.

As shown in Fig. 5 when the "Follow-Up Time" feature has a value of 32 for a specific instance, there is 0.08 probability that the instance will be predicted as belonging to the "survival" class, when the "Anemia" feature has a value of 0, there is 0.07 probability that the instance will be predicted as belonging to the

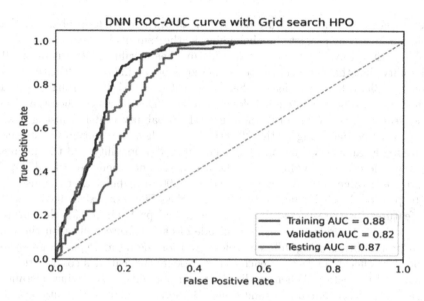

Fig. 4. ROC-AUC curve DNN model with Grid Search HPO and after SMOTE

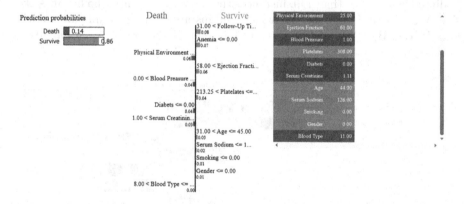

Fig. 5. XGBoost Model Explainability using LIME.

"survival" class, when the "Ejection Fraction" feature has a value between 58 and 61, there is 0.06 probability that the instance will be predicted as belonging to the "survival" class, when the "Platelets" feature has a value between 213 and 308, there is 0.04 probability that the instance will be predicted as belonging to the "survival" class, when the "Age" feature has a value between 31 and 45, there is 0.03 probability that the instance will be predicted as belonging to the "survival" class, when the "Serum Sodium" feature has a value less than or equal to 126, there is 0.02 probability that the instance will be predicted as belonging to the "survival" class, when the "Smoking" feature has a value of 0, there is 0.01 probability that the instance will be predicted as belonging to the "survival" class, when the "Gender" feature has a value of 0, there is 0.01 probability that the instance will be predicted as belonging to the "survival" class.

When the "Physical Environment" feature has a value of 25, there is 0.06 probability that the instance will be predicted as belonging to the "Death" class, when the "Blood Pressure" feature has a value of 1, there is 0.04 probability that the instance will be predicted as belonging to the "Death" class, when the "Diabetes" feature has a value of 0, there is 0.04 probability that the instance will be predicted as belonging to the "Death" class, when the "Serum Creatinine" feature has a value between 1 and 1.11, there is 0.03 probability that the instance will be predicted as belonging to the "Death" class, and when the "Blood Type" feature has a value between 8 and 11, there is 0.00 probability that the instance will be predicted as belonging to the "Death" class Fig. 6 shows the most important features for the selected instance's predicted probabilities for the survive class. It shows the direction and magnitude of each feature's effect on the predicted probabilities, using different colors (red for the negative effect, green for the positive effect). The size of each bar represents the feature's contribution to the survival prediction. When predicting the probability of surviving, instances with an age greater than 60, instances with Diabetes equal to zero (contains a not diabetic patient), instances with a hypertensive patient, instances with Serum Sodium level greater than 140, instances that contain patients who live in Ankesha, Ayo, Azena, Banja and Bahir Dar, and instances that contain patients who have B, B+, B−, AB, and AB+ tend to have lower the predicted probability of Survive class.

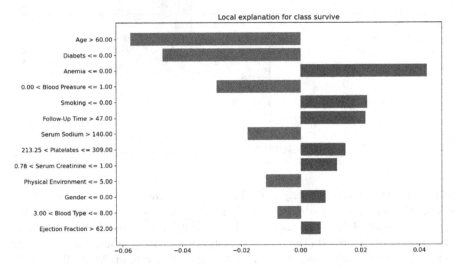

Fig. 6. Contribution of each feature for Survive class using LIME. (Color figure online)

Instances with patients who haven't Anemia, instances with patients who don't smoke, instances with patients whose Follow-up Time is greater than 47, instances with patients who have a level of Platelets between 213 and 309 in their blood, instances with patients whose level of Serum Creatinine is between 0.78 and 1, instances that contains female patients and instances with patients Ejection Fraction greater than 62 tend to have increase the predicted probability of Survive class.

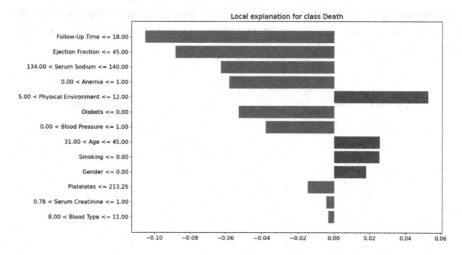

Fig. 7. Contribution of each feature for Death class using LIME. (Color figure online)

Figure 7 shows the most important features for the selected instance's predicted probabilities for the Death class. It shows the direction and magnitude of each feature's effect on the predicted probabilities, using different colors (red for the negative effect, green for the positive effect). The size of each bar represents the feature's contribution to the Death prediction. When predicting the probability of Death, instances with patients with Follow-up Time less or equal to 18, instances with patients with Ejection Fraction less than or equal to 45, instances with patients with Serum Sodium level between 134 and 140, instances with anemic patients, instances with not diabetic patients, instances with hypertensive patients, instances with patients who have a level of Platelets less than or equal to 213, instances with a level of Serum Creatinine between 0.78 and 1, and instances with AB−, O, and O+ patients' Blood Type tend to have lower the predicted probability of Death Class.

Instances that contain patients who live in Chagni, Damat, Dangila, Debub Achefer, Ebnat, Estie and Fagita Lekoma, instances with an age between 31 and 45, instances with patients who don't smoke, and instances that contains female patients tend to have increased the predicted probability of Death class.

Using SHAP

SHAP is a methodology that can be employed to understand the results of machine learning models and pinpoint their advantages and deficiencies. SHAP

provides a method to assign importance scores to feature variables to understand their impact on the model's predictions. Figure 8 waterfall plot shows the
significance of each characteristic to the final prediction of the survival class,
with the features listed in descending order of impact. Positive contributions
(in red) increase the prediction, while negative contributions (in blue) decrease
it. Zero values in the SHAP waterfall plot represent features that do not have
any impact on the prediction. This means that those particular features have no
contribution to either increasing or decreasing the predicted value and therefore
their SHAP values are zero.

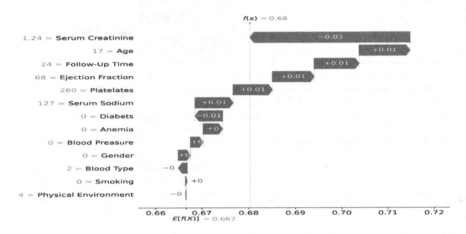

Fig. 8. The contribution of each feature for the Survive class using SHAP. (Color figure
online)

Features such as Age, Follow-up Time, Ejection Fraction, Platelets and Serum
Sodium have a strong positive impact on survive class by increasing the probability of Survive. While Serum Creatinine, and Diabetes, have a negative impact
on survival, have decrease the probability of survive. Anemia, Blood Pressure,
Gender, Blood Type, Smoking and Physical Environment have no impact on
survival prediction. The magnitudes indicate the strength of the impact while
the direction (positive or negative) indicates the influence on survival. Figure 9
waterfall plot shows the significance of each characteristic to the final prediction
of the Death class, with the features listed in descending order of impact. Positive contributions (in red) increase the prediction, while negative contributions
(in blue) decrease it. Zero values in the SHAP waterfall plot represent features
that do not have any impact the prediction. This means that those particular
features have no contribution to either increasing or decreasing the predicted
value and therefore their SHAP values are zero.

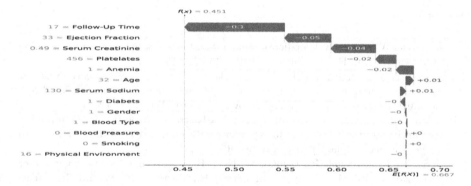

Fig. 9. The contribution of each feature for Death class using SHAP. (Color figure online)

Features such as Follow-up Time, Ejection Fraction, Serum Creatinine, Platelets and Anemia have a strong negative impact on death, and have a decrease probability of death. While 'Age', and 'Serum Sodium', have a positive impact on Death, have increased the probability of death. Diabetes, Gender, Blood Type, Gender, Blood Type, Blood Pressure, Smoking, and Physical Environment have no impact on survival prediction.

6   Conclusion

Cardiovascular diseases (CVDs) refer to various ailments affecting the blood vessels and heart. These illnesses include heart failure, heart attack, stroke, rheumatic heart disease, and hypertension. Heart failure is becoming an increasingly significant worldwide public health issue and a primary cause of death.

This study aims to create a machine learning model that can predict if a heart failure patient will survive, while also being able to provide an explanation for its predictions. We have used DT, LR, KNN, DNN, and XGBoost machine learning algorithms for prediction. We apply these algorithms before and after the SMOTE balancing and with hyper-parameter optimization techniques. Experimental analyses have shown that XGBoost outperformed other models with an accuracy of 86.5%, precision of 88%, sensitivity of 86%, F1 score of 86%, and 0.93 AUC curve value with Grid Search after SMOTE.

Acknowledgments. We want to express our gratitude to Injibara general hospital and Felege Hiwot referral hospital workers.

References

1. Almazroi, A.A.: Survival prediction among heart patients using machine learning techniques. Math. Biosci. Eng. **19**(1), 134–145 (2022)
2. Angaw, D.A., Ali, R., Tadele, A., Shumet, S.: The prevalence of cardiovascular disease in Ethiopia: a systematic review and meta-analysis of institutional and community-based studies. BMC Cardiovasc. Disord. **21**(1), 1–9 (2021)
3. Aslan, M.F., Sabanci, K., Durdu, A.: A CNN-based novel solution for determining the survival status of heart failure patients with clinical record data: numeric to image. Biomed. Signal Process. Control **68**, 102716 (2021)
4. Bergstra, J., Bardenet, R., Bengio, Y., Kégl, B.: Algorithms for hyper-parameter optimization. In: Shawe-Taylor, J., Zemel, R., Bartlett, P., Pereira, F., Weinberger, K. (eds.) Advances in Neural Information Processing Systems, vol. 24. Curran Associates, Inc. (2011). https://proceedings.neurips.cc/paper_files/paper/2011/file/86e8f7ab32cfd12577bc2619bc635690-Paper.pdf
5. Brochu, E., Cora, V.M., De Freitas, N.: A tutorial on Bayesian optimization of expensive cost functions, with application to active user modeling and hierarchical reinforcement learning. arXiv preprint arXiv:1012.2599 (2010)
6. Buda, M., Maki, A., Mazurowski, M.A.: A systematic study of the class imbalance problem in convolutional neural networks. Neural Netw. **106**, 249–259 (2018)
7. Chandrashekar, G., Sahin, F.: A survey on feature selection methods. Comput. Electr. Eng. **40**(1), 16–28 (2014)
8. Chicco, D., Jurman, G.: Machine learning can predict survival of patients with heart failure from serum creatinine and ejection fraction alone. BMC Med. Inform. Decis. Mak. **20**(1), 1–16 (2020)
9. Foziljonova, N., Wasito, I.: Prediction of survival rate of heart failure patients using machine learning techniques. J. Theor. Appl. Inf. Technol. **100**(9), 2703–2714 (2022)
10. Guo, X., Yin, Y., Dong, C., Yang, G., Zhou, G.: On the class imbalance problem. In: 2008 Fourth International Conference on Natural Computation, vol. 4, pp. 192–201. IEEE (2008)
11. Herrera, D., et al.: Association between periodontal diseases and cardiovascular diseases, diabetes and respiratory diseases: consensus report of the Joint Workshop by the European Federation of Periodontology (EFP) and the European Arm of the World Organization of Family Doctors (WONCA Europe). J. Clin. Periodontol. **50**(6), 819–841 (2023)
12. Hussen, N.M., Workie, D.L., Biresaw, H.B.: Survival time to complications of congestive heart failure patients at Felege Hiwot comprehensive specialized referral hospital, Bahir Dar, Ethiopia. PLoS One **17**(10), e0276440 (2022)
13. Ishaq, A., et al.: Improving the prediction of heart failure patients' survival using smote and effective data mining techniques. IEEE Access **9**, 39707–39716 (2021)
14. Kebede, B., Getachew, M., Molla, Y., Bahiru, B., Dessie, B.: Management, survival, and predictors of mortality among hospitalized heart failure patients at Debre Markos comprehensive specialized hospital, Northwest Ethiopia: prospective cohort study. SAGE Open Med. **9**, 20503121211057336 (2021)
15. Maharana, K., Mondal, S., Nemade, B.: A review: data pre-processing and data augmentation techniques. Global Transitions Proc. **3**(1), 91–99 (2022)
16. Moreno-Sanchez, P.A.: Development of an explainable prediction model of heart failure survival by using ensemble trees. In: 2020 IEEE International Conference on Big Data (Big Data), pp. 4902–4910. IEEE (2020)

17. Newaz, A., Ahmed, N., Haq, F.S.: Survival prediction of heart failure patients using machine learning techniques. Inform. Med. Unlocked **26**, 100772 (2021)
18. Obaid, H.S., Dheyab, S.A., Sabry, S.S.: The impact of data pre-processing techniques and dimensionality reduction on the accuracy of machine learning. In: 2019 9th Annual Information Technology, Electromechanical Engineering and Microelectronics Conference (IEMECON), pp. 279–283. IEEE (2019)
19. Özbay Karakuş, M., Er, O.: A comparative study on prediction of survival event of heart failure patients using machine learning algorithms. Neural Comput. Appl. **34**(16), 13895–13908 (2022)
20. Probst, P., Boulesteix, A., Bischl, B.: Tunability: importance of hyperparameters of machine learning algorithms. J. Mach. Learn. Res. **20**, 53:1–53:32 (2018). https://api.semanticscholar.org/CorpusID:88515435
21. Schaer, R., Müller, H., Depeursinge, A.: Optimized distributed hyperparameter search and simulation for lung texture classification in CT using Hadoop. J. Imaging **2**(2), 19 (2016)
22. Stiglic, G., Kocbek, P., Fijacko, N., Zitnik, M., Verbert, K., Cilar, L.: Interpretability of machine learning-based prediction models in healthcare. Wiley Interdisc. Rev. Data Min. Knowl. Discov. **10**(5), e1379 (2020)
23. Tefera, T.A., Muleta, G., Tadesse, K.: Bayesian survival analysis of heart failure patients: a case study in Jimma University Medical Center, Jimma, Ethiopia (2021)
24. Wistuba, M., Schilling, N., Schmidt-Thieme, L.: Hyperparameter optimization machines. In: 2016 IEEE International Conference on Data Science and Advanced Analytics (DSAA), pp. 41–50. IEEE (2016)
25. Yibre, A.M., Koçer, B.: Semen quality predictive model using feed forwarded neural network trained by learning-based artificial algae algorithm. Eng. Sci. Technol. Int. J. **24**(2), 310–318 (2021)

Natural Language Processing, Text and Speech Processing

Transfer of Models and Resources for Under-Resourced Languages Semantic Role Labeling

Yesuf Mohamed[1]([envelope])[iD] and Wolfgang Menzel[2][iD]

[1] Information Science, Addis Ababa University, Addis Ababa, Ethiopia
`yesuf.mohamed@aau.edu.et`
[2] Computer Science, Hamburg University, Hamburg, Germany
`wolfgang.menzel@uni-hamburg.de`

Abstract. Identifying and labeling the semantic roles of words and phrases within a sentence is a crucial task in the field of natural language processing (NLP), known as semantic role labeling (SRL). Although SRL techniques have been well-developed for languages with high resources, under-resourced languages pose substantial challenges due to the scarcity of annotated data and language-specific tools. This paper investigates the potential of utilizing models and resources from languages with high resources to enhance SRL performance in under-resourced languages. The paper provides an overview of the SRL process, identifies the specific challenges faced by under-resourced languages, and proposes techniques for transferring models and resources from languages with high resources.

We investigate a potential solution, which is based on a transfer learning approach that utilizes a pre-trained English SRL model. Our approach leverages a substantial parallel English-Amharic corpus to build the Amharic SRL dataset without the need for extensive manual annotation. Our model achieves an outstanding F1 score of 85.2% on the Amharic test set, showcasing its effectiveness in identifying and labeling semantic roles in Amharic sentences. Using transfer learning and extensive parallel corpora, our method overcomes data scarcity and proves the possibility of creating advanced language processing tools for under-resourced languages.

Keywords: Natural language processing (NLP) · Cross-lingual Transfer learning · Semantic role labeling (SRL) · Domain adaptation · Under-resourced languages

1 Introduction

Semantic Role Labeling (SRL) is a natural language processing (NLP) task that aims to identify the underlying meaning or semantic structure of a sentence by identifying the relationships between its constituents. Specifically, SRL involves assigning specific semantic roles to each word or phrase in a sentence, such as agent (the doer of an action), patient (the entity that undergoes the action), instrument (the tool or means used to perform an action), and location (the place where an action occurs) [1]. SRL can be used in many NLP applications, such as question answering, information retrieval, and machine translation, as it helps to disambiguate sentence meaning and improve accuracy [2]. Generally, SRL answers the question "Who did What to Whom, How, When, and Where" in sentences [3]. The lack of annotated data and the linguistic differences between high-resourced and under-resourced languages pose significant challenges to this task. However, leveraging well-developed models and resources from other languages, language processing tasks, or language domains to improve SRL accuracy in under-resourced languages is an active area of research. Nowadays, one of the methods used to resolve this issue is transfer learning [4, 15].

Transfer learning involves, leveraging a model learned from a high-resourced language or domain to improve the performance of a model in an under-resourced language [5]. This approach is particularly useful in scenarios where there is a scarcity of labeled data in the target language, which makes it challenging to train high-performing models. One common transfer learning method for Semantic Role Labeling (SRL) is to use models trained on large-scale labeled datasets in a source language and adapt them to a target language [15]. This paper will discuss the transfer of models and resources from high-resourced to under-resourced languages for SRL. We will also mention the challenges and limitations of building automatic semantic role labeling systems for an under-resourced language, and answer the question of how these challenges can be addressed.

2   Overview of Semantic Role Labeling for Under-Resourced Languages

2.1   Explanation of the Semantic Role Labeling Process

The goal of SRL is to identify the relationships between words or phrases in a sentence and the actions they describe [6]. By assigning semantic roles, SRL disambiguates sentence meaning and facilitates NLP applications like machine translation, information retrieval, and question answering [7]. The process of semantic role labeling typically involves two steps. Firstly, it identifies the main predicate or verb in the sentence, which represents the action being performed. Once the main predicate has been identified, the next step finds the arguments of the predicate, namely the words or phrases that play specific roles in the action described by the predicate [8, 15]. Several types of semantic roles can be assigned to words or phrases, including the agent, which is the entity that performs the action; the patient, which is the entity that is affected by the action; the instrument, which is the tool or means used to perform the action; and the location, which is the place where the action occurs [1, 15].

The semantic role labeling process can be performed manually, or automatically. Manual annotation requires human annotators to identify the semantic roles in a sentence, which can be time-consuming and expensive. On the other hand, automatic semantic role labeling involves training a machine learning model on annotated data, which can then be used to automatically assign semantic roles to new words or phrases in a sentence [15].

2.2 Challenge in Automatic Semantic Role Labeling of Under-Resourced Languages

The main challenge in developing automatic SRL systems for under-resourced languages is the lack of annotated data. In general, the accuracy of automatic semantic role labeling systems highly depends on the amount and quality of annotated data available for training the model. However, for many under-resourced languages, annotated data are scarce, making it difficult to develop an accurate semantic role labeling system [15].

2.3 Approaches for Leveraging Models and Resources from High-Resourced Languages or Domains

There are various approaches for leveraging models and resources of highly developed languages or domains to improve the performance of under-resourced languages/domains semantic role labeling (SRL). These approaches include:

Cross-Lingual Transfer Learning: This approach involves using pre-trained models in high-resourced languages and adapting them to under-resourced languages. The pre-trained model is fine-tuned using a small amount of labeled data in the under-resourced language, and the patterns learned from the high-resourced language are transferred to the under-resourced language. This approach has shown promising results in improving SRL accuracy for under-resourced languages [15]. Cross-lingual transfer learning has become possible due to advances in machine learning and natural language processing (NLP) in recent years. One key development has been the availability of large amounts of data in high-resourced languages, which allows models to be trained on a large number of examples. Advances in data availability, as well as techniques to map annotations from a source to a target language, have all contributed to the feasibility of transfer learning [15].

Multi-task Learning: In this approach, SRL is jointly learned with another related task such as part-of-speech tagging or named entity recognition. By learning multiple related tasks simultaneously, the model learns shared representations that can improve the performance of SRL in under-resourced languages [11]. However multi-task learning requires considerable amounts of annotated data for all the tasks which are included. Thus, it might reduce the necessary corpus size while spending more effort on multi-task annotation [15].

Domain Adaptation: Domain adaptation is an approach for adapting a model trained on one domain (e.g., news articles) to another domain (e.g., scientific papers). Domain adaptation can be useful for SRL because labeled data can be scarce or expensive in some domains, and it may be more efficient to use a pre-trained model and adapt it to the target domain than to train a new model from scratch. This could help to improve the performance of the SRL system on the new domain without requiring a large amount of labeled data [15].

Data selection is another way to perform domain adaptation. If only a small corpus of domain-specific data is available, but a larger one with out-of-domain data, it is possible to select the most similar data items from the out-of-domain corpus to extend the in-domain collection [15].

2.4   Evaluation Metrics of Different Approaches for Semantic Role Labeling

To evaluate the quality of SRL results, the following evaluation metrics are used [9,10] individually or in combination. They can also be applied to measure the effectiveness of transfer learning in SRL systems.

Precision: Precision measures the percentage of correctly identified semantic roles out of all the roles predicted by the SRL system. A high precision score indicates that the SRL system is accurate in identifying the semantic roles [15].

Recall: Recall measures the percentage of correctly identified semantic roles out of all the roles present in the text. A high recall score indicates that the SRL system is effective in identifying all the semantic roles in the text [15].

F1-Score: The F1-score is the harmonic mean of precision and recall. It provides a balanced measure of both precision and recall and is often used as a single evaluation metric for SRL systems [15].

Error Rate: The error rate measures the percentage of incorrectly identified semantic roles out of all the roles predicted by the SRL system. A low error rate indicates that the SRL system is accurate in identifying the semantic roles [15].

2.5   Case Studies and Examples

To demonstrate the approaches for SRL in under-resourced languages, we will discuss four specific case studies.

Ilseyar Alimova et al. [4] apply cross-lingual transfer learning to improve SRL accuracy for Russian. They use a semantic role labeling model pre-trained on a high-resourced language, English, and a Russian FrameBank dataset for evaluation. The results show that the transfer learning approach improved the SRL accuracy from 78.4% F1 score to 79.8% in Russian, achieving a good performance [15].

The study of Fariz Ikhwantri et al. [12] uses multi-task learning with SRL as a primary task and entity recognition as a secondary one. They trained their model with only 6057 annotated sentences and achieved an F1 score of 86.26% by using multi-task learning methods [15].

To overcome the dataset scarcity for multi-lingual conversational semantic role labeling, which is the extension of SRL that is specifically designed for handling the context of a conversation, the study by Wu et al. [13] applied the cross-lingual method. They achieved an F1 score of 62.67% by fine-tuning the model parameters [15].

To improve the performance of the automatic Semantic Role Labeling task for Portuguese, Sofia Oliveira et al. [5], explored a model architecture that uses a pre-trained Transformer-based model. They also leverage cross-lingual transfer learning using multilingual pre-trained models and transfer learning from dependency parsing in Portuguese to further improve the results. The authors were able to achieve a substantial improvement in the state-of-the-art performance for Portuguese by over 15 F1 score [15].

3   Advantages and Challenges of Using Rich Resources for Semantic Role Labeling of an Under-Resourced Language

3.1   Advantages

Improved Performance: The use of well-developed models or resources from alternative sources can significantly improve the performance of SRL in under-resourced languages. This is because high-resourced language/domains have more annotated data, which can be used to train SRL models with higher accuracy [15].

Reduced Annotation Cost: The use of resources and models from high-resourced languages/domains can also reduce the cost of annotation in under-resourced languages or tasks. Instead of annotating data from scratch, transfer learning approaches can be used to adapt well-trained models to under-resourced languages. This can significantly reduce the amount of labeled data needed for training [15].

Faster Development: The use of resources from high-resourced languages can also speed up the development of SRL systems for under-resourced languages. Instead of starting from scratch, pre-trained models and tools can be used as a starting point for developing SRL systems for under-resourced languages [15].

3.2   Challenges

The main challenges of using rich models and datasets to train a semantic role labeling system for an under-resourced language are:

Linguistic Differences: One major challenge to transferring resources and models from high-resourced languages is the existence of linguistic differences between the languages involved. For example, languages with different word order, and morphological complexity can affect the performance of SRL models trained on high-resourced languages. High-resourced languages such as English have grammatical features that might not be present in under-resourced languages. For instance, Amharic has different grammatical features than English, if we take subject-verb agreement, for example, an Amharic verb can carry more information than an English verb. A single Amharic verb can carry three kinds of information (number, gender, and tense) at the same time. An English verb can't carry gender information at all, even if it can carry two (number and tense) information at the same time, it is restricted to third person singular. Therefore, directly applying a semantic role labeling system trained on an English dataset to Amharic would not yield accurate results [15].

Domain Differences: Another challenge is the existence of domain differences between high-resourced and under-resourced language datasets [3]. The rich resources may be in a different domain, which may not be relevant to the under-resourced language. This can lead to poor performance of the SRL system in the target domain or language. Even if domain adaptation is recommended to overcome the dataset scarcity of under-resourced languages, it is not easy to accomplish [15].

Data Availability: The availability of annotated data in the under-resourced language can be a problem. While transfer learning can reduce the amount of labeled data needed to reach the same level of accuracy, some labeled data is still required. If there is a lack of annotated data in the under-resourced language, it can be difficult to develop an accurate SRL system [15].

3.3   Possible Solution

One possible solution to meet the challenges of using rich datasets for training a semantic role labeling system for an under-resourced language is adapting the dataset to the target language or task. This involves developing a mapping between the semantic roles in the high-resourced language and the target language or task [15].

4   Methodology

In this study, we detail the methodology used for the development of an Amharic Semantic Role Labeling (SRL) model using transfer learning. Our goal was to overcome the challenge of limited labeled Amharic data by leveraging a pre-trained English SRL model and a collection of "parallel corpora for Ethiopian languages" [14], which contains a substantial number of corpora in English, Amharic, and other Ethiopian local languages. Figure 1 shows the proposed architecture to transfer knowledge and resources from a high-resourced language model and to generate SRL-labeled Amharic sentences without having labeled datasets.

The numbers in circles on top of the architecture components show the order of the process in creating both Amharic SRL-labeled texts and the fine-tuned Amharic SRL-labeler model. The following section describes the architecture.

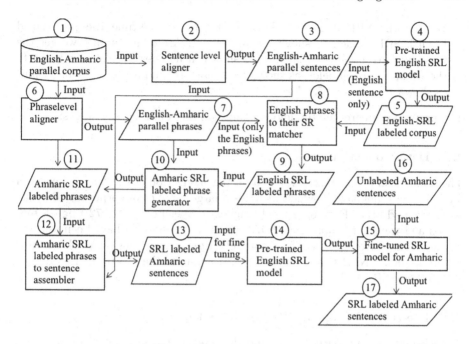

Fig. 1. Proposed architecture.

- English-Amharic parallel corpus in number 1 is the input for both sentence-level and phrase-level aligners. The sentence-level aligner generates English-Amharic parallel sentences, as shown in number 3, while the phrase-level aligner generates English-Amharic parallel phrases, as shown in number 7.
- The pre-trained English SRL model in number 4 receives only the English sentence from the parallel sentence in number 3 as input and produces SRL-labeled English sentences as output, shown in number 5.
- English-phrase-to-their-SRLmatcher in number 8 receives two inputs: English SRL-labeled corpus from number 5 and unlabeled English-Amharic phrases from number 7. It then produces English SRL-labeled phrases, as shown in number 9.
- Amharic-SRL-labeled-phrase-generator in number 10 receives two inputs: English SRL-labeled phrases in number 9 and English-Amharic parallel phrases in number 7. It then produces Amharic SRL-labeled phrases. After obtaining both labeled and unlabeled English phrases, it transfers the labels from labeled to unlabeled English phrases and then again transfers the labels to equivalent Amharic phrases.
- Amharic-SRL-labeled-phrases-to-sentence-assembler in number 12 takes Amharic SRL-labeled phrases and English-Amharic parallel sentences as input. It then assembles the Amharic SRL phrases into Amharic SRL-labeled sentences. The English-Amharic parallel sentence is used to refer to the original Amharic sentences and build the same sentences but SRL-labeled.

– We use the SRL-labeled Amharic sentences to fine-tune the pre-trained English SRL model, as shown in number 14. After fine-tuning, we get the fine-tuned Amharic SRL labeler, as shown in number 15. We give input to the Amharic SRL labeler, as shown in number 16, and it produces an output, as shown in number 17.

The proposed research methodology comprises the following key stages:

4.1   Data Collection

The foundation of this research relied on the acquisition of the "parallel corpora for Ethiopian languages", which includes a large number of corpora in English, Amharic, and other Ethiopian local languages. Specifically, 27,972 English sentences and their corresponding 27,972 Amharic translations have been selected from this extensive corpora. These parallel sentence pairs provide a rich resource for cross-lingual NLP tasks, allowing us to fine-tune the pre-trained model for the Amharic SRL task.

4.2   Pre-trained English SRL Model Selection

To kickstart our approach, we downloaded the pre-trained semantic role labeling model from AllenNLP [16], a well-established NLP toolkit. This pre-trained model for English SRL offered a strong foundation for capturing linguistic features and structures. We harnessed its power to adapt and extend its capabilities to Amharic.

4.3   Alignment and Amharic SRL-Labeled Corpus Generation

The selected corpus consists of parallel English and Amharic texts. Since it is not separated at a sentence or paragraph level, it is difficult to identify the boundary of the translation on the other side. So, alignment between the two languages is crucial. The process began at the sentence level using Moses alignment tool [17], establishing correspondences between the two languages. This alignment tool facilitated a one-to-one component between English and Amharic texts, the components are sentences in sentence level alignment and words in word level alignment. The alignment ensures that each English sentence has an aligned Amharic counterpart. The following table shows samples of sentence pairs from the sentence-aligned parallel corpus (Table 1).

The English part of the aligned sentences was input into a pre-trained English SRL model, which automatically assigned semantic roles to the English words within each sentence. This step allowed us to generate SRL-labeled English sentences, detailing the roles of words in the context of the sentence's semantics without the need for manual annotation (Table 2).

With SRL-labeled English sentences at our disposal, we proceeded to the phrase-level alignment. The parallel sentences were now aligned at the phrase level, meaning that each English phrase was aligned with its corresponding

Table 1. Samples of sentence-level aligned parallel corpus

English sentence	Equivalent Amharic sentence
The generous person will be blessed, For he shares his food with the poor.	ለጋስ ሰው ይባረካል፤ ምግቡን ለድሃ ያካፍላልና
Next they departed from Makhieloth and camped at Tahath	ቀጥሎም ከማቅሄሎት ተነስተው በታሃት ሰፈሩ
And he placed his hands on them and departed from there.	እጁንም ከጫነባቸው ቧላ ከዚያ ስፍራ ተነስቶ ሄደ
After they got up into the boat, the wind-storm abated.	ጀልባው ላይ ከወጡ ቧላ አውሎ ነፋሱ ጸጥ አለ

Table 2. Sample of given input and generated output by the pre-trained model

Given English sentence	Produced English sentence with SRL tags
The generous person will be blessed, For he shares his food with the poor	(The generous person)/ARG1 will/ARGM-MOD be blessed/V, (For he/ARG0 shares/V (his food)/ARG1 (with the poor)ARG2) ARGM-CAU
Next they departed from Makheloth and camped at Tahath	Next/ARGM-TMP they/ARGO departed/V (from Makheloth)/ARG1 and camped/V at Tahath/ARGM-LOC
And he placed his hands on them and departed from there	And/ARGM-DIS he/ARG-0 placed/V (his hands)/ARG1 (on them)/ARG2 and (departed)/V (from there)/ARG1
After they got up into the boat, the windstorm abated	(After they/ARG0 got/V up/ARGM-DIR (into the boat)/ARG2)/ARGM-TMP, (the windstorm)/ARG1 abated/V

Amharic phrase. This alignment ensured that we had one-to-one correspondence between the English and Amharic phrases (Table 3).

In the next step, we systematically transferred the English role labels to their corresponding Amharic counterparts. This process allowed us to associate semantic roles with the Amharic phrases, effectively generating SRL-labeled Amharic phrases. This methodology enabled us to develop an Amharic SRL model without the need for extensive manual annotation of Amharic data. Table 4 shows how the transfer of labels from English to Amharic was performed.

4.4 Amharic SRL Model Fine-Tuning

The final phase was the fine-tuning of the pre-trained English SRL model with the newly created Amharic SRL-labeled corpus. This process enabled the model to adapt to the Amharic language's unique linguistic structures and semantics. We evaluated the model's performance using the above-mentioned metrics, precision, recall, and F1-score.

Table 3. Samples of phrase-level-aligned English-Amharic phrases

English phrases	Equivalent Amharic phrases
The generous person	ለጋስ ሰው
will be blessed	ይባረካል
For he shares	ያካፍላል
his food	ምግቡን
with the poor.	ለድሃ
Next	ቀጥሎም
they departed	ተነስተው
from Makheloth	ከማቅሄሎት

Table 4. Label transfer from English to Amharic equivalents.

English phrases	Labels	Amharic phrases
The generous person	ARG1	ለጋስ ሰው
will be blessed	V	ይባረካል
For he shares	he/ARG0 shares/V	ያካፍላል
his food	ARG1	ምግቡን
with the poor.	ARG2	ለድሃ
Next	ARGM-TMP	ቀጥሎም
they departed	they/ARG0 departed/V	ተነስተው
from Makheloth	ARG1	ከማቅሄሎት

5   Results and Discussion

Our approach achieved an F1 score of 85.2%. This shows that it has success-fully combined the information from the pre-trained model and the parallel English-Amharic corpus, providing a bridge for knowledge transfer from a rich resource (pre-trained English model) to a resource-scarce language (Amharic). The method significantly reduced the need for costly and time-consuming man-ual annotation of Amharic data, making it a cost-effective and efficient strategy for NLP tasks in under-resourced languages.

The results of our research demonstrate the efficacy of the transfer learn-ing approach in developing an Amharic SRL model without the dependency on labeled Amharic data. The fine-tuned pre-trained English SRL model exhibited promising performance in the Amharic language, as evidenced by the evaluation metrics.

The performance suggests that this transfer learning approach has the potential to be extended to other languages with limited resources, opening doors to new possibilities in the field of cross-lingual NLP. This research not only contributes to the development of NLP tools for Amharic but also offers a valuable methodology for advancing NLP in other resource-constrained languages.

6 Conclusion

This study has explored the feasibility and effectiveness of transfer learning in the domain of Amharic Semantic Role Labeling (SRL). By leveraging the rich resources available in high-resourced languages, we have addressed the substantial challenge of limited annotated data in the Amharic language. Our research demonstrates the promising potential of transfer learning to bridge the linguistic and resource gaps, making NLP tasks more accessible and efficient for under-resourced languages like Amharic. The methodology detailed in this study showcases a systematic approach for developing an Amharic SRL model without the need for extensive manual annotation. It offers a cost-effective and scalable solution that significantly reduces the time and resources required for developing NLP tools in resource-scarce languages.

The results of our research indicate that the transfer learning approach, particularly the adaptation of pre-trained English SRL models to the Amharic language, yields substantial improvements in under-resourced language SRL accuracy. The fine-tuning process successfully aligns the model with the unique linguistic structures and semantics of Amharic, paving the way for broader applications in the field of cross-lingual NLP.

As the importance of NLP for under-resourced languages continues to grow, the methodology presented in this study opens new avenues for the development of NLP tools in resource-constrained environments. While we have focused on Amharic, this approach can be extended to other under-resourced languages, further advancing the field of cross-lingual NLP.

7 Future Work

The journey of NLP in under-resourced languages is ongoing, and future work may delve deeper into optimizations, explore additional linguistic resources, and tackle domain-specific challenges to enhance the performance of Amharic SRL models. To improve the model's generalizability and prevent overfitting, it is crucial to increase the size and diversity of the parallel dataset used for fine-tuning. This could involve incorporating data from various domains, such as news articles, social media posts, and literary works.

References

1. Gildea, D., Jurafsky, D.: Automatic labeling of semantic roles. Comput. Linguist. **28**(3), 245–288 (2002). https://doi.org/10.1162/089120102760275983
2. Sequeira, J., Gonçalves, T., Quaresma, P.: Semantic role labeling for Portuguese–a preliminary approach–. In: Caseli, H., Villavicencio, A., Teixeira, A., Perdigão, F. (eds.) PROPOR 2012. LNCS (LNAI), vol. 7243, pp. 193–203. Springer, Heidelberg (2012). https://doi.org/10.1007/978-3-642-28885-2_22
3. Do, Q.N.T., Bethard, S., Moens, M.-F.: Facing the most difficult case of semantic role labeling: a collaboration of word embeddings and co-training. In: Proceedings of COLING 2016, the 26th International Conference on Computational Linguistics: Technical Papers, pp. 1275–1284. The COLING 2016 Organizing Committee, Osaka, Japan (2016). https://aclanthology.org/C16-1121
4. Alimova, I., Tutubalina, E., Kirillovich, A.: Cross-lingual transfer learning for semantic role labeling in Russian. In: Proceedings of the 4th International Conference on Computational Linguistics in Bulgaria (CLIB 2020), pp. 72–80. Department of Computational Linguistics, IBL - BAS, Sofia, Bulgaria (2020). https://aclanthology.org/2020.clib-1.8
5. Oliveira, S.A., Loureiro, D., Jorge, A.M.: Improving Portuguese semantic role labeling with transformers and transfer learning. In: 2021 IEEE 8th International Conference on Data Science and Advanced Analytics (DSAA), pp. 1–9 (2021)
6. Gormley, M., Mitchell, M., Durme, B., Dredze, M.: Low-resource semantic role labeling, vol. 1, pp. 1177–1187 (2014). https://doi.org/10.3115/v1/P14-1111
7. Hailu, B., Assabie, Y., Sinshaw, Y.: Semantic role labeling for Amharic text using multiple embeddings and deep neural network. IEEE Access **11**, 33274–33295 (2023). https://doi.org/10.1109/ACCESS.2023.3263147
8. Larionov, D., Shelmanov, A., Chistova, E., Smirnov, I.: Semantic role labeling with pretrained language models for known and unknown predicates. In: Proceedings of the International Conference on Recent Advances in Natural Language Processing (RANLP 2019), pp. 619–628. INCOMA Ltd., Varna, Bulgaria (2019). https://doi.org/10.26615/978-954-452-056-4073, https://aclanthology.org/R19-1073
9. Arora, A., Malireddi, H., Bauer, D., Sayeed, A., Marton, Y.: Multi-task learning for joint semantic role and proto-role labeling (2022)
10. Márquez, L., Carreras, X., Litkowski, K.C., Stevenson, S.: Special issue introduction: semantic role labeling: an introduction to the special issue. Comput. Linguist. **34**(2), 145–159 (2008). https://doi.org/10.1162/coli.2008.34.2.145
11. Kozhevnikov, M., Titov, I.: Cross-lingual transfer of semantic role labeling models. In: Proceedings of the 51st Annual Meeting of the Association for Computational Linguistics (Volume 1: Long Papers), pp. 1190–1200. Association for Computational Linguistics, Sofia, Bulgaria (2013). https://aclanthology.org/, P13-1117
12. Ikhwantri, F., et al.: Multi-task active learning for neural semantic role labeling on low resource conversational corpus. In: Proceedings of the Workshop on Deep Learning Approaches for Low-Resource NLP, pp. 43–50. Association for Computational Linguistics, Melbourne (2018). https://doi.org/10.18653/v1/W18-3406, https://aclanthology.org/W18-3406
13. Wu, H., Tan, H., Xu, K., Liu, S., Wu, L., Song, L.: Zero-shot cross-lingual conversational semantic role labeling. In: Findings of the Association for Computational Linguistics: NAACL 2022, pp. 269–281. Association for Computational Linguistics, Seattle, United States (2022). https://doi.org/10.18653/v1/2022.findings-naacl.20, https://aclanthology.org/2022.findings-naacl.20

14. Abate, S.T., et al.: Parallel corpora for bi-lingual English-Ethiopian languages statistical machine translation. In: Proceedings of the 27th International Conference on Computational Linguistics, pp. 3102–3111. Association for Computational Linguistics, Santa Fe, New Mexico, USA (2018). https://aclanthology.org/C18-1262
15. Yesuf Mohamed, W.M.: Can the rich help the poor? Transfer of knowledge and resources for under resourced languages semantic role labeling. EasyChair (2023). https://easychair.org/publications/preprintopen/GJbh
16. Gardner, M., et al.: AllenNLP: a deep semantic natural language processing platform. In: Park, E.L., Hagiwara, M., Milajevs, D., Tan, L. (eds.) Proceedings of Workshop for NLP Open Source Software (NLP-OSS), pp. 1–6. Association for Computational Linguistics, Melbourne, Australia (2018). https://doi.org/10.18653/v1/W18-2501, https://aclanthology.org/W18-2501
17. Koehn, P., et al.: Moses: open source toolkit for statistical machine translation (2007)

Speaker Identification Under Noisy Conditions Using Hybrid Deep Learning Model

Wondimu Lambamo[1(✉)] ⓘ, Ramasamy Srinivasagan[1,2] ⓘ, and Worku Jifara[1] ⓘ

[1] Adama Science and Technology University, 1888 Adama, Ethiopia
wondimuwcu@gmail.com
[2] King Faisal University, Al Hofuf 31982, Al-Ahsa, Saudi Arabia
rsamy@kfu.edu.sa

Abstract. Speaker identification is a biometric mechanism that determines a person who is speaking from a set of known speakers. It has vital applications in areas like security, surveillance, forensic investigations, and others. The accuracy of speaker identification systems was good by using clean speech. However, the speaker identification system performance gets degraded under noisy and mismatched conditions. Recently, a network of hybrid convolutional neural networks (CNN) and enhanced recurrent neural network (RNN) variants have performed better in speech recognition, image classification, and other pattern recognition. Moreover, cochleogram features have shown better accuracy in speech and speaker recognition under noisy conditions. However, there is no attempt conducted in speaker recognition using hybrid CNN and enhanced RNN variants with the cochleogram input to enhance the models' accuracy in noisy environments. This study proposes a speaker identification for noisy conditions using a hybrid CNN and bidirectional gated recurrent unit (BiGRU) network on the cochleogram input. The models were evaluated by using the VoxCeleb1 speech dataset with real-world noise, white Gaussian noises (WGN), and without additive noise. Real-world noises and WGN were added to the dataset at the signal-to-noise ratio (SNR) of -5 dB up to 20 dB with 5 dB intervals. The proposed model attained an accuracy of 93.15%, 97.55%, and 98.60% on the dataset with real-world noises at SNR of -5 dB, 10 dB, and 20 dB, respectively. The proposed model shows approximately similar performance on both real-world noise and WGN at similar SNR levels. Using the dataset without additive noise the model achieved 98.85% accuracy. The evaluation accuracy and the comparison with the previous works indicate that our model has better accuracy.

Keywords: Speaker Identification · Convolutional Neural Network · Cochleogram · Bidirectional Gated Recurrent Unit · Real-World Noises

1 Introduction

Speaker recognition [1] is a biometric technology that uses the characteristics of speech to classify or identify a person from a set of known speakers.

T. G. Debelee et al. (Eds.): PanAfriConAI 2023, CCIS 2068, pp. 154–175, 2024.
https://doi.org/10.1007/978-3-031-57624-9_9

Two different individuals cannot produce the same type of speech in any circumstance because of unique physiological and behavioral characteristics [2]. Each person's speech production system is unique in shape, size and other structures which helps to produce unique speech. In addition, every person has a unique way of speaking style, timing between two words, choice of word, and so on [3]. Human-to-computer interaction for various services can be easy and natural using the systems that use speech. Speech-based systems are easily accessible and acceptable by the users and implementation requires resources of low cost. These advantages increased the demand for systems that use speech in various application areas. The speaker recognition is one of the speech-based systems which has enormous applications in the real world. It can be implemented from small to large organizations. For instance, it is crucial in security [4], authentication, forensics investigation [5], video conference, surveillance [6] and financial transactions [7].

Based on how the claimed speakers are classified from the set of speakers, we can classify speaker recognition as identification vs. verification [8]. Speaker verification systems determine whether the person speaking to the system exists or not in the trained speaker database. In speaker verification, there are only two states of decision that either accept the speaker or reject [9]. The likelihood ratio of test utterances is computed from each speaker class. Then the speaker with the maximum likelihood ratio is selected and compared with the threshold values to make a decision. Speaker verification is an open-set classification where samples are assumed to come from known speakers or unknown speakers. Speaker identification is selected in our study.

Speaker identification automatically classifies an individual who gives an utterance from a registered or trained set of speakers in the model [10]. The system determines the speaker as the one who has the maximum similarity ratio. In speaker identification, the likelihood ratio of the test sample is computed from each class trained in the model. The person who claimed with the utterance can be decided as the speaker class of the maximum likelihood ratio. Identification is also known as closed set recognition, in which each claimed utterance is expected to be from known speakers. Both verification and identification follow a sequence of steps such as speech pre-processing, feature extraction, training and classification. Features extraction techniques and classification models are important components that have critical importance in the performance of the systems [11]. Classical machine learning [12] and deep neural networks [13] are commonly employed classification models in speaker identification. In this study, deep learning models were employed for speaker identification in noisy conditions.

The speaker identification system accuracy was good by using clean speech or under a clean environment and without mismatch between the samples. However, the systems' effectiveness in speaker recognition gets degraded by various types of environmental noises, channel variations, languages, and physical and behavioral changes in the speaker. These factors are the main challenges for implementing speaker identification systems in various real-world applications.

Numerous research works were conducted to enhance the efficiency of speaker recognition under noisy conditions. Most of the previous works in speaker recognition have been conducted by employing classical machine learning approaches.

Recently, deep neural networks outperformed the previous machine learning approaches in speech analysis including speaker identification. Convolutional neural networks (CNN) and recurrent neural networks (RNN) [14], were commonly employed deep neural models for speaker identifications. CNN is one of the deep neural networks that use a feed-forward interconnection of layers without memory to store the previous state and no cycle. It automatically extracts training parameters from the input and learns adaptively from each parameter. It has advantages in extracting the dependency of features for the short term and requires low computational resources because of a limited number of parameters [15]. RNN has memory to store the information of long-term feature dependency and has been commonly employed for time series data analysis. Standard RNN can be affected by gradient vanishing and exploding when there is long-term dependency occurs between the features.

Enhanced variants of RNN which are long-short term memory (LSTM) [16] and gated recurrent unit [17], were employed in various studies to handle the limitations of standard RNN. The GRU, which is an advanced LSTM version has an advantage in extracting and learning one direction long term correlation between features sequentially [18]. Compared with LSTM, GRU models have fewer parameters which minimizes the computation cost of the models. However, the GRU model misses the other direction correlation information of the feature because it only finds one direction correlation information. Bidirectional gated recurrent unit (BiGRU) extracts the long-term dependency of features of the two directions which handles the limitation of GRU. Recently, combinations of CNN and RNN variants have been showing promising performance in most of the application areas. However, most of the speaker identification methods employed either CNN or variants of RNN models. In our study, a hybrid CNN and BiGRU network is employed for speaker identification.

Deep learning models use different types of features which is generated at different stages of speech analysis for speaker identification. Commonly used features include Mel frequency cepstral coefficient (MFCC), gamma tone frequency cepstral coefficient (GFCC), spectrograms, raw waveform, Mel spectrograms, and cochleogram. Spectrograms that were generated from speech were mostly used in speaker identification studies using deep neural networks. However, the speaker identification performance of the spectrograms gets degraded under noisy and mismatched conditions. The Cochleogram of the speech performed better than the spectrogram in speaker recognition in noisy and mismatched situations. However, none of the hybrid CNN and enhanced RNN variants employed cochleogram input to enhance the speaker identification accuracy of the models in noisy and mismatched conditions.

This study proposes a speaker identification using a hybrid CNN and BiGRU for noisy conditions. The models use cochleogram as the input. The models were evaluated by using the VoxCeleb1 speech dataset with real-world noise,

WGN and without additive noises. The real-world and WGN were added to each utterance with a signal-to-noise ratio (SNR) of $-5\,$dB to $20\,$dB in intervals of $5\,$dB. The proposed model accuracy was evaluated concerning the performance of other models such as two-dimensional CNN (2DCNN), CNN-LSTM, CNN-BiLSTM and CNN-GRU to present the models' effectiveness. Comparisons with the previous works were also conducted to indicate the models' effectiveness.

The remaining sections of this paper are structured as follows: Sect. 2 discusses related works. Section 3 presents the methodology which includes the cochleogram generation process, CNN, BiGRU, and proposed model architecture. Section 4 discusses experiments which include datasets, experimental setup, and results and discussion. Section 5 concludes the study and forwards future works.

2   Related Works

Speaker recognition systems have a vital role in real-world applications. It can be implemented in small to large organizations for different kinds of applications such as speaker authorization, authentication, surveillance, classification and segmentation. Therefore, many researchers, organizations and individuals are interested in speaker recognition and a large number of research works have been conducted in the area. Previous studies conducted in the area widely used the classical machine learning approach. Some of the classical machine learning approaches include the Gaussian mixture model (GMM) [19], support vector machine (SVM) [20], and Hidden Markov Model (HMM) [21]. These approaches have been using handcrafted features in speech analysis including speaker recognition. For example, in the studies [22] and [23], speaker verification systems were conducted using i-vector and GMM classifiers with MFCC respectively. The speaker recognition proposed in the study [24], has used a GMM machine learning approach and MFCC features of the utterances. Speaker recognition conducted with the GMM approach on the MFCC feature was proposed as the biometrics for home device control [25] and remote identification over VoIP [26]. In the reference [27], the GMM method together with the MFCC and inverse MFCC feature extraction techniques were employed in speaker recognition.

The performance of the speaker recognition systems using MFCC features was good using clean speech and without mismatch between the samples. However, the performance of the MFCC features gets degraded with real-world noises, background noises, and changes in the physical and behavioral characteristics of speakers [28]. In the research works [29] and [30], GFCC features have shown better performance over the MFCC features under noisy conditions in speaker recognition. Therefore, speaker recognition systems using conventional machine learning methods under noisy and mismatched conditions widely employed GFCC. For instance, the speaker identification proposed in the study [31], employed the GMM approach and GFCC on noisy acoustic datasets. In the reference [32], a speaker identification developed for forensic applications in noisy environments applied the GMM-UBM and GFCC features. Speaker verification in real-world noise existing conditions proposed in the study [33], employed the model GMM and GFCC features.

Deep neural networks performed better than classical machine learning approaches in various areas including speaker recognition. Deep learning models have been employing various types of input for speaker recognition. Some studies in speaker recognition employed images of the features that were commonly used in conventional machine learning. In the study [34] and [35], the CNN model which is one of the deep neural networks was employed in speaker identification together with the handcrafted feature MFCC. In the reference [36], a visual geometry group (VGG) and the MFCC features were proposed for speaker identification. A Siamese network of CNN and MFCC features was employed in the speaker verification model for cross-device platforms [37]. GFCC features were used together with the CNN model to develop a speaker identification model under noisy conditions [38]. Deep learning models together with MFCC and GFCC features have achieved better performance than classical classifiers in speaker recognition. However, MFCC and GFCC features were not as effective as other inputs like spectrograms for speaker recognition using deep neural networks. In the study [39], spectrograms have achieved superior performance than MFCC in speaker recognition using the deep neural network model CNN. In the reference [40], spectrogram features have shown better accuracy than raw speech signal and MFCC in speaker recognition using deep machine learning. This is because spectrogram features are rich in acoustic features to characterize the speaker which helps the deep learning networks easily learn the correlation between the features. Like the systems that use MFCC features, the performance of the speaker identification model which uses a spectrogram gets degraded in noisy conditions. In the study [41], we analyzed the noise robustness of cochleogram and spectrogram in speaker recognition by using the CNN model's different kinds of architectures. In the analysis, cochleogram outperformed the spectrogram in speaker recognition under different ratios of noises in the speech and both features have approximately similar performance in clean environments.

Recently, combinations of the CNN model with the variants of the enhanced RNN model have been achieving better performance in speech recognition, image classification, time series data analysis and other natural language processing. In the studies [42] and [43], speaker identification and verification models conducted using a combination of CNN and LSTM models have enhanced the performance over the previous models. Reference [44], also employed a combination of CNN and BiLSTM models together with the spectrogram of speech input for language identification and the results have shown improvement in existing works. The audio-visual recognition model proposed for biometric application in the study [45], also indicated the effectiveness of a hybrid network of CNN and BiLSTM models. In the study [17], a hybrid CNN and GRU network were employed for speaker identification with the spectrogram of Aishell-1 datasets.

In the above discussions, most of the research works conducted in the area of speaker recognition have employed either the CNN or RNN variants model.

A limited number of studies have been conducted in speaker recognition areas using combinations of deep neural networks especially CNN and enhanced RNN variants. The RNN variants that have been employed in hybrid models have limitations. For example, LSTM has the disadvantage of overfitting and high computation costs because of more number of gates. The GRU network model is an improved LSTM version, which has fewer gates than LSTM. The GRU models extract one-direction long-term correlation information of the features that miss the other-direction correlation information. Bidirectional GRU (BiGRU) models were proposed to extract and learn both direction (i.e., forward and backward) correlation information of the features that handle the limitations of GRU. Moreover, most of the speaker recognition models developed using deep neural networks commonly used spectrogram input. However, there is no attempt conducted in speaker identification using a combination of models CNN and BiGRU network to increase the accuracy of models. In addition, none of the hybrid models of CNN and RNN variants used cochleogram input for speaker recognition to increase the models' performance in noisy conditions. This study proposes a speaker identification using a combination of CNN and BiGRU models on the cochleogram input to improve the effectiveness in noisy conditions. Our model integrated the advantages of the models CNN and BiGRU with the advantages of cochleogram input. CNN components extract short-term correlations of the features and learn adaptively from fewer parameters. The BiGRU component extracts long-term feature dependency in the forward and backward directions sequentially. It also handles gradient vanishing problems and converges faster during the model training. Cochleogram features have a finer resolution for the low-frequency speech sample and rich in acoustics features of the speaker and consist of noise information in advance.

3   Methodology

This study proposed a combination of CNN and BiGRU models on the cochleogram input for speaker identification in a noisy environment. Our model consists of three basic components such as cochleogram generation, CNN, and BiGRU components. The theoretical background of each component and how each component has been employed in the proposed model is discussed in the following subsections in detail.

3.1   Cochleogram Generation

A Cochleogram is a representation of speech in a two-dimensional (2D) time-frequency image to employ in speech analysis tasks. The time and frequency of the speech were represented in the x-axis and y-axis of the cochleogram. The color in the cochleogram image represents the amplitude of the sample. The cochleogram generation process comprises pre-emphasis of input speech, framing speech into segments, applying a window function to each frame, computing fast Fourier transform of each frame, gamma tone filter bank, and computing power spectrum from gamma tone filters which is shown in Fig. 1.

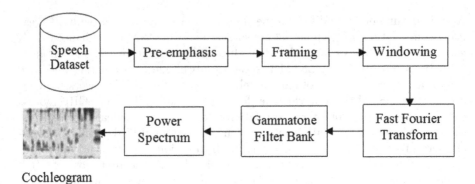

Cochleogram

Fig. 1. Cochleogram generation process

In the speech samples lower frequency component always dominates the higher frequency component, which may reduce the effectiveness of the systems in noisy environments. Pre-emphasis is useful to compensate for the energy in high-frequency speech samples concerning energy in the low-frequency samples. For sample s(n) and emphasis factor f, pre-emphasized sample P(n) can be calculated according to Eq. 1.

$$P(n) = s(n) - f * s(n-1) \tag{1}$$

The speech is very dynamic and obtaining stable information from a long speech is difficult. Segmenting speech into the frames enables us to get stable acoustic information. The recommended frame duration in various speech analysis tasks ranges from 20 ms to 50 ms with an overlap of 30% to 50% between the frames. Applying a window function to each frame helps to reduce discontinuity between the frames. Generally, the hamming window is commonly used in speech analysis for different types of applications. Each frame F(n) can be windowed as shown in Eq. 2 by using the window function W(n) is calculated as shown in Eq. 3.

$$H(n) = F(n) * W(n) \tag{2}$$

$$W(n) = 0.5 - 0.46 \cos(2n\pi/(M-1)) \tag{3}$$

Obtaining important features from the time domain of the speech signal is difficult. During speech analysis tasks signals are converted from the time domain into the frequency domain. Fast Fourier Transform (FFT) computes the frequency domain of the samples from the time domain, the result is known as a spectrum or periodogram of the sample. FFT can be computed from the input frame H(n) and N number of points according to Eq. 4.

$$T(k) = \sum_{n=0}^{N-1} H(n)e^{-2j\pi nk/N} \tag{4}$$

A Cochleogram can be obtained from the gamma tone filters in the equal rectangular bandwidth (ERB) scale. ERB measures the psychoacoustics of the speech to determine the estimated bandwidths of human hearing filters. Gamma tone filters simulate the human auditory system which helps to model the speaker, language, speech and other information from the speech. It has advantages in representing speech of low frequency in finer resolution. For the FFT of the speech frames with amplitude a and time t, the gamma tone filter of filter order n and phase shift delta can be computed according to Eq. 5.

$$g(t) = at^{n-1}e^{-2\pi b_m t}\cos(\pi f_c t + \delta) \tag{5}$$

Central frequency f_c for the mth gammatone filter is computed from the low-frequency f_L, high-frequency f_H and constant value q = 1000/4.37 according to Eq. 6.

$$f_c = (-q) + (f_H + q)\exp(m/M(-\ln(f_H + q) + \ln(f_L + q))) \tag{6}$$

The Equal Rectangular Bandwidth (ERB) scale of the central frequency f_c can be calculated as in Eq. 7.

$$ERB(f_c) = 24.7(4.37 * (f_c/1000) + 1) \tag{7}$$

The bandwidth b_m of the gammatone filters can be calculated as shown in Eq. 8.

$$b_m = 1.019 * ERB(f_c) \tag{8}$$

Each gamma tone filter can be used as a feature to model the speaker, speech, language, and emotion that exists in a speech. The power spectrum of the gamma tone filters is stacked together to form cochleogram of the given speech. Finally, the proposed model takes the cochleogram as input for speaker identification.

3.2 Convolutional Neural Network

CNN model is one of the deep neural networks that is composed of a feed-forward network of layers without cycle and memory. CNN models extract features automatically from the input data without human intervention. Various architectures of the CNN model were widely applied in areas like medical image classification, computer vision, biometric recognition, speech recognition, and so on. CNN models represent each class with few parameters, which saves computational resources and it has strong learning ability from few parameters. It comprises different types of layers that are interconnected to each other for feature extraction, learning the pattern, and classifying. The basic layers in CNN are convolution, pooling and fully connected, layers. In the convolution layer, useful parameters to train the model will be extracted by using the filters. The models can be trained by the parameters of the filters throughout the training. Pooling layers are commonly applied for the sub-sampling of the feature maps. It can be employed following the convolution layers to minimize the dimension of the features for the next layers. There are several pooling methods such as tree, gated, average, min, max, global average and so on. Max, min, and global average pooling were mostly employed in various deep neural network-based applications.

3.3   Bidirectional Gated Recurrent Unit

Recurrent Neural Network (RNN) is a type of deep neural network that has been broadly adopted in time series data analysis. It has a memory to store information about the previous sequences and forward the important state of the previous state to the next layers based on the gate's decision. RNN models extract the correlation between the features sequentially in a time series. However, standard RNN models can be affected by gradient exploding and vanishing in the features that have a long sequence of dependency. Standard RNN may be unable to extract the long-term correlation of the features because of its gradient vanishing and exploding problem.

Enhanced variants of RNN were commonly used in time series data analysis and other applications because of their advantages in handling the limitations of standard RNN. They store long-term dependency information of the features that help to handle gradient vanishing and exploding. LSTM and GRU are the two widely applied enhanced variants of RNN in speaker recognition and other areas. In LSTM network models there are three types of gates such as input, forget, and output gates. In the input gate, the information that should be stored in the long-term memory is decided. The information that should be discarded from the long-term memory is decided by the forget gate. New short-term memory information gets generated from the current input, previous short-term memory and newly generated long-term memory on the output gate. LSTM models are more complex because they consist of more gates and a larger number of parameters than GRU. An overfitting problem and high computational resource consumption are the disadvantages of LSTM networks.

GRU is one of the variants of RNN which uses gates to regulate, manage and manipulate information in the cells of the neural network. GRU models have only two gates (i.e., Update and reset gates) that can handle the limitation of LSTM models. In GRU, the update gate shows the combinations of the input gate and forget gate of LSTM models. The update gate of GRU models mainly controls the amount of information that should pass from the earlier state to the next. The update gate minimizes the risk of gradient vanishing by storing the information of the previous state of long sequence dependency. In the reset gate of the GRU models, the amount of previous information that should be discarded is decided (i.e., whether the previous cell state is useful or not). GRU architectures are simple in structure because it has only two types of gates in each of the cell. Since GRU networks extract the correlation of features sequentially in one direction only, it miss the dependency information in the other direction.

Bidirectional gated recurrent units (BiGRU) have been recommended for extracting the backward and forward long-term dependency information of the features sequentially. The BiGRU layer is the interconnection of GRU layers in forward and backward mechanisms to extract both directions' long-term correlation information of the features. Moreover, BiGRU networks converge after a few iterations during the model evaluation.

3.4 Proposed Model Architecture

The proposed speaker identification is composed of CNN and BiGRU models combination which employ a cochleogram of speech as an input. There are three basic components in our model which are Cochleogram generation, CNN layers and BiGRU layers. A Cochleogram generation module has been employed to convert each of the utterances in the dataset into the cochleogram. The CNN component was employed to automatically extract the short-term dependency of the speaker features. The BiGRU component was applied to extract the forward and backward long-term feature dependency from the parameters sequentially. The architecture of the CNN-BiGRU model for speaker identification is shown in Fig. 2.

Cochleogram of the speech has been used as an input for training, validating and testing the models. Therefore, each of the utterances in the selected dataset has been converted into cochleogram to feed the model. Cochleogram generation have been conducted based on the sequence of tasks in Fig. 1. First, the speech was pre-emphasized with an emphasis factor of 0.97, then it was framed into segments of 30 ms with 10 ms overlap and the hamming window was applied. From each frame, the fast Fourier transform with the filters of 128 and filter points of 2048 were computed. To compute gamma tone filters from each FFT, the lower frequency was set to 0 Hz and the higher frequency of 8000 Hz was used. The central frequency of the gamma tone filter was computed from the lower and higher frequencies according to Eq. 6 and its ERB scale was computed based on Eq. 7. The bandwidth of the gamma tone filters was computed from the ERB scale of central frequency according to Eq. 8. Gamma tone filters were computed from each of the FFTs based on central frequency, bandwidth, time, amplitude, and filter order according to Eq. 5. Each of the gamma tone filter's power spectrum was stacked together to generate cochleogram of the utterance. The cochleogram of size $224 \times 224 \times 3$ was generated which was an important shape for the CNN input tensor of the proposed model. Various forms of speech representations for analysis are illustrated in Fig. 3. In Fig. 3(a) speech is represented in raw waveform and Fig. 3(b) speech is represented in cochleogram.

The proposed model comprises two convolution layers preceding two BiGRU layers. Two of the convolution layers were employed consecutively at the beginning of the model to extract short-term spatial feature correlation and learn adaptively from the parameters. The kernel size of 3×3, the same padding, ReLu activation and normal kernel initializer were employed in each of the convolution layers. Each convolution layer was followed by the max-pooling layer of pool size 2×2 to reduce the feature dimensions for the next layers. Batch normalization was also employed after each max-pooling layer to normalize the features. The first convolution layer has the filter size of 16 filters and the second convolution layer has the filter size of 32 filters.

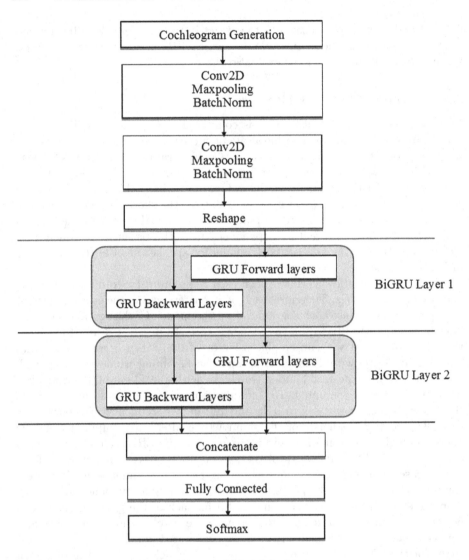

Fig. 2. Proposed Model Architecture

The BiGRU layers were employed after the second max-pooling layer of the proposed model. Each BiGRU layer takes the second max-pooling layer's output by reshaping it into the appropriate shape. The max-pooling output shape is not similar to the BiGRU layer's input shape. The reshape layer was applied in between the second max-pooling and BiGRU layers, it mainly converts the parameters into the appropriate input shape of the BiGRU layer. The BiGRU layers have 256 cell units for parameter extraction and training of the model. The outputs of both BiGRU layers were concatenated to feed the fully connected layer. The concatenated output was used as an input for a fully connected layer

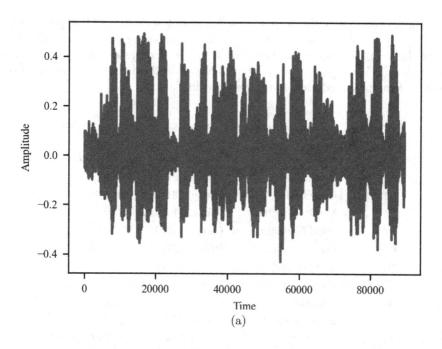

(a)

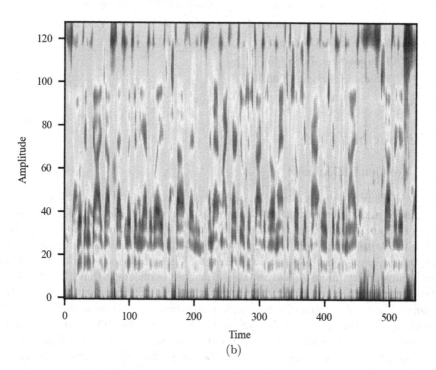

(b)

Fig. 3. Speech representation; (a) Raw waveform and (b) Cochleogram

to train the networks of the proposed model. The classification of the speaker into the respective classes was conducted by using the softmax activation function. Since the VoxCeleb1 speech dataset has 1251 speaker classes, we set the activation function class equal to the number of classes in the dataset. Implementation details of the proposed model are presented in Table 1.

Table 1. Summary of CNN-BiGRU model for proposed model

Layer No.	Layer Name	Output Shape	Parameter
01	InputLayer	(None, 224, 224, 3)	0
02	Convolution	(None, 224, 224, 16)	448
03	MaxPooling2D	(None, 112, 112, 16)	0
04	BatchNormalization	(None, 112, 112, 16)	64
05	Convolution	(None, 112, 112, 32)	4640
06	MaxPooling2D	(None, 56, 56, 32)	0
07	BatchNormalization	(None, 56, 56, 32)	128
08	Reshape	(None, 56, 1792)	0
09	Bidirectional GRU	(None, 512)	3148800
10	Bidirectional GRU	(None, 512)	3148800
11	Concatenate	(None, 1024)	0
12	Fully Connected	(None, 1251)	1282275
13	Activation	(None, 1251)	0

Total params: 7,585,155

Trainable params: 7,585,155

4   Experiment

4.1   VoxCeleb1 Dataset

The VoxCeleb1 speech dataset has been used to evaluate the performance of our model and other additional deep neural networks. The VoxCeleb1 speech dataset was primarily prepared for speech analysis applications including speaker recognition. It is a free and publicly available dataset that helps the researcher to conduct various speech analysis experiments. The dataset consists of more than one hundred fifty thousand utterances that were collected from 1251 persons. The celebrity videos on YouTube were the main source for each speech in the dataset. The total number of male speakers and female speakers in the VoxCeleb1 speech dataset is approximately 55% and 45% respectively. The utterances with different types of English language accents were included in the dataset. Since each utterance was obtained from celebrity videos recorded in the real world, it contains various types of speaking styles, age groups, language backgrounds, and professions. The sample frequency in the utterances of the dataset is 16000 Hz. In each speaker class of the dataset, the total number of utterances and their length varies. Primarily, the original split (i.e., dev and test split) of the dataset was first merged into one to split for our experiment. In our experiment, the dataset was split into 80% for training, 10% for validation and 10% for testing. Although the VoxCeleb1 dataset contains noises in various ratios, it was assumed as a clean dataset. The models were evaluated by using the dataset with real-world noises, WGN and without additive noises. The dataset with the real-world noise was generated by adding randomly selected real-world noise to each utterance in the dataset. The real-world noises employed during the experiment included babble, restaurant, and street noises. The dataset with the WGN was generated by adding the noise generated by Python code to each utterance. Both real-world noises and WGN were added to each utterance at the SNR of −5 dB to 20 dB in the interval of 5 dB.

4.2 Experimental Setup

Since the proposed model contains a combination of deep neural networks (i.e., CNN and BiGRU models), the TensorFlow library was used to experiment. TensorFlow library is a Python language-based library recommended for deep learning model development purposes. The management of important packages, deployment of a deep learning model and other speech analysis tasks were employed in the anaconda navigator. We have customized the Spafe package (simplified Python audio features extraction) to convert utterances into cochleograms. The cochleogram of utterances was used as an input for both training and testing each deep learning model. The hardware NVIDIA TITAN Xp GPU was used to conduct the experiments. The hardware environment we used for the experiment was GPU computers which is more effective than CPU computers for deep neural network models. During the experiment, each model used an initial batch size of 32. The root mean square propagation (RMSprop) was used as the optimizer to obtain the optimum results of the model. Moreover, in each model, the loss function known as categorical cross-entropy was employed to calculate the loss. The model's training was conducted for 50 epochs. In each of the epochs, the accuracy and loss of both training and validation were calculated.

4.3 Results and Discussion

In this study, the speaker identification accuracy of the 2DCNN, CNN-LSTM, CNN-BiLSTM, CNN-GRU, and CNN-BiGRU models was evaluated using the input cochleogram of utterances on the VoxCeleb1 dataset at different noise ratios. From the models, CNN-BiGRU was proposed for speaker identification under noisy conditions. Each of the models was evaluated by the dataset with real-world noise, WGN and without additive noises. The evaluation of each model for real-world noise and WGN was conducted at the SNR of −5 dB to 20 dB in intervals of 5 dB. The evaluation of the models at different levels of noise ratio was repeated for 10 rounds. The average of the results in each round experiment is used to report the overall performance of models. To present the model's performance graphically, one training progress of the models on the dataset with real-world noise at SNR of 20 dB was selected. The accuracy and loss of each model on the dataset with the real-world noise at SNR of 20 dB are shown in Fig. 4. (a) and (b), respectively. The results in Fig. 4(a) show that the 2DCNN achieved the least accuracy and CNN-BiGRU has the highest accuracy at each of the epochs. Moreover, CNN-BiGRU converges faster and has relatively consistent accuracy than other models. Other models CNN-LSTM, CNN-BiLSTM and CNN-GRU have better accuracy than 2DCNN but lower accuracy than CNN-BiGRU. The 2DCNN, CNN-LSTM, and CNN-BiLSTM models converge much slower than CNN-GRU and CNN-BiGRU. The results in Fig. 4b also indicated that 2DCNN has more loss than other models. The loss of the CNN-BiGRU model is smaller than the loss of other models.

Table 2, presents an average speaker identification accuracy of the models on the dataset with WGN. Each model's performance was evaluated at different levels of SNR (i.e., at SNR = −5 dB, 0 dB, 5 dB, 10 dB, 15 dB and 20 dB). The average accuracy of the models on the dataset with a very high WGN ratio (i.e., at SNR = −5 dB) ranges from 71.66% to 93.63%. The average accuracy of the models on the dataset with a similar ratio of WGN and signal (i.e., SNR = 0 dB) ranges from 83.7% to 96.24%. The performance of the models on the dataset with WGN at SNR = 5 dB ranges from 86.61% to 97.41%. At medium WGN ratio in the signals of the dataset (i.e., at SNR = 10 dB), the average accuracy of the models ranges from 87.48% to 97.83%. At SNR equals 15 dB and 20 dB the average accuracy of the models ranges from 88.90% to 98.22% and 90.78% to 98.73% respectively. The results show that 2DCNN achieved the lowest accuracy which ranges from 71.66% to 90.78% on the dataset with WGN at SNR ranging from −5 dB to 20 dB, respectively. From the models used in this study, the CNN-BiGRU model achieved the highest accuracy on the dataset with WGN at different ratio of SNR which ranges from 93.63% to 98.73% at SNR from −5 dB to 20 dB, respectively. At SNR equals −5 dB, the CNN-BiGRU model has shown improvements of 1.8% to 21% on other models. At SNR equals 20 dB the CNN-BiGRU model achieved an improvement of 0.3% to 7.95% accuracy. Generally, the speaker identification performance of the CNN-BiGRU model with cochleogram input is better than other models on the dataset with WGN at different levels of SNR.

Table 3, reports an average speaker identification accuracy of models on the dataset with the real-world noise. Each model's speaker identification accuracy is presented at different levels of the SNR (i.e., at SNR = −5 dB, 0 dB, 5 dB, 10 dB, 15 dB, and 20 dB). At low SNR the accuracy of each model is lower than their performance at high SNR. For example, the accuracy of 2DCNN, CNN-LSTM, CNN-BiLSTM, CNN-GRU, and CNN-BiGRU at SNR = −5 dB is 71.48%, 86.39%, 87.38%, 91.33% and 93.15%, respectively which is lower than their performance at SNR = 0 dB and above. At each of the SNR levels, 2DCNN achieved the lowest accuracy than others. The CNN-BiGRU model has shown

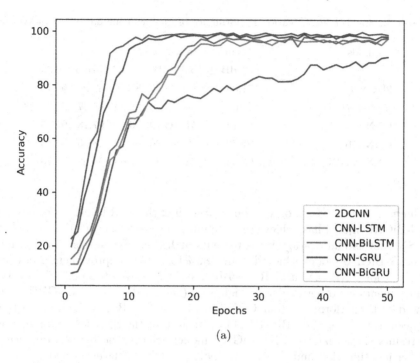

(a)

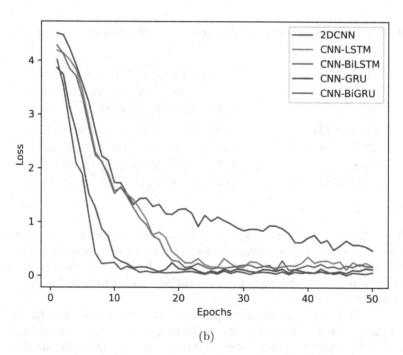

(b)

Fig. 4. Speaker identification performance of models using dataset with real-world noise at $SNR = 20\,dB$; (a) Accuracy and (b) Loss

Table 2. Speaker Identification Performance of models using dataset with WGN

Methods	Accuracy (in %) at SNR					
	−5 dB	0 dB	5 dB	10 dB	15 dB	20 dB
2DCNN	71.66	83.7	86.61	87.48	88.90	90.78
CNN-LSTM	86.61	89.85	93.86	96.35	97.07	97.36
CNN-BiLSTM	87.58	91.51	94.72	96.44	97.18	97.55
CNN-GRU	91.82	94.22	95.96	97.52	98.05	98.43
CNN-BiGRU (Proposed)	93.63	96.24	97.41	97.83	98.22	98.73

the highest accuracy than other models at each of the SNR levels. Moreover, the CNN-BiGRU model has achieved maximum improvement over other models at low SNR levels and average improvements at higher SNR levels. For example, the CNN-BiGRU model achieved from 1.82% to 21.67% improvement over other models at low SNR (i.e. at SNR = −5 dB), and it achieved from 0.36% to 7.98% improvement over other models at high SNR (i.e., 20 dB). The CNN-GRU model showed better performance than CNN-LSTM and CNN-BiLSTM, but lower than the accuracy of the CNN-BiGRU model at each of the SNR levels. In general, the results show that the CNN-BiGRU model achieved better performance in speaker identification under real-world noise at different levels of SNR.

Table 3. Speaker identification performance of models using dataset with real world noises

Methods	Accuracy (in %) at SNR					
	−5 dB	0 dB	5 dB	10 dB	15 dB	20 dB
2DCNN	71.48	83.52	86.42	87.28	88.71	90.62
CNN-LSTM	86.39	89.63	93.63	96.11	96.84	97.14
CNN-BiLSTM	87.38	91.31	94.51	96.22	96.97	97.35
CNN-GRU	91.33	94.03	95.76	97.31	97.85	98.24
CNN-BiGRU (Proposed)	93.15	95.86	96.37	97.55	98.07	98.60

Table 4, presents the speaker identification performance of the models on the VoxCeleb1 dataset without additive noise. The results show that the performance of the models ranges from 91.52% to 98.58%. 2DCNN has the lowest average accuracy which is 91.52% and the CNN-BiGRU model has the highest average accuracy which is 98.85%.

Table 5, provides comparisons of proposed model performance with the existing works to illustrate the efficiency of the proposed model. The existing works developed by using the deep learning model on the VoxCeleb1 dataset were selected for comparison. The results of the study [46] proposed by Nagrani et al., the study [47] proposed by Kim and Park, and the study [48] proposed by Ding

Table 4. Speaker Identification Performance of Models on the Dataset without Additive Noise

Method	Accuracy (in %)
2DCNN	91.52
CNN-LSTM	97.44
CNN-BiLSTM	97.79
CNN-GRU	98.52
CNN-BiGRU (Proposed)	98.85

et al. were selected for comparison. The comparison shows that the proposed model has superior accuracy than the existing works. The speaker identification accuracy of the study on [46, 47], and [48] are 92.10%, 95.30% and 96.01% respectively. The proposed model in this study achieved an average accuracy of 98.85%, which outperformed the existing works in speaker identification.

Table 5. Comparison of proposed model performance with existing works

Method	Dataset	Accuracy (in %)
Nagrani et al. [46]	VoxCeleb1	92.10
Kim and Park [47]	VoxCeleb1	95.30
Ding et al. [48]	VoxCeleb1	96.01
CNN-BiGRU (Proposed)	VoxCeleb1	98.85

Generally, our model has achieved better accuracy than other models evaluated in this study at different levels of SNR. Our model has also achieved superior accuracy than other models on the dataset without additive noises. The comparison also confirmed that our model has better accuracy than previous works used as a baseline. The main reason for the enhancement of the accuracy is that the model has integrated the advantages of 2DCNN and BiGRU networks. Moreover, using cochleogram of the speech as an input improved the efficiency of the models under noisy conditions.

5   Conclusion

In this paper, a hybrid deep neural network is proposed for speaker identification in noisy conditions. The model was constructed from a combination of CNN and BiGRU models. A Cochleogram of the utterances was employed as the model input. The model integrated the advantages of the two deep neural networks (i.e., CNN and BiGRU models) and the cochleogram features. Our model and other deep neural networks used in this paper were evaluated on the VoxCeleb1 speech dataset with and without additive noises. Both real-world and WGN noise were added to the dataset at the SNR of $-5\,$dB to $20\,$dB in the interval of $5\,$dB. The results show that the proposed model has higher accuracy than other models on the datasets with real-world noise, WGN and without additive noise. At low SNR levels, the proposed model has shown maximum improvement over the other models and average improvements at higher SNR levels. The accuracy of the proposed model on the dataset with real-world noise and WGN is approximately similar at similar SNR levels. On the dataset without additive noise, the proposed model also achieved the highest accuracy among others. The comparisons with existing works also indicate that the proposed model has better performance. The main reason for the performance improvements is that the model integrated the advantages of the CNN, BiGRU and Cochleogram features. In the future, further studies can be conducted by using a fusion of features, enhancing the input of the models, and hybrid models of either CNN and RNN variants or other models to enhance the speaker identification performance under noisy and mismatched environments.

References

1. Kinnunen, T., Li, H.: An overview of text independent speaker recognition from features to super-vectors. Speech Commun. **52**(1), 12–40 (2010)
2. Ahmed, S., Mamun, N., Hossain, A.: Cochleagram based speaker identification using noise adapted CNN. In: 5th International Conference on Electrical Engineering, Information and Communication Technology (ICEEICT) (2021)
3. Gustavo, A.: Modeling prosodic differences for speaker recognition. Speech Commun. **49**(4), 77–291 (2007)
4. Selvan, K., Joseph, A., Babu, A.: Speaker recognition system for security applications. In: IEEE Recent Advances in Intelligent Computational Systems (RAICS), Trivandrum, India (2013)
5. Han, K., Omar, M., Pelecanos, J., Pendus, C., Yaman, S., Zhu, W.: Forensically inspired approaches to automatic speaker recognition. In: IEEE International Conference on Acoustics, Speech and Signal Processing (ICASSP), Prague, Czech Republic (2011)
6. Alegre, F., Soldi, G., Evans, N., Fauve, B., Liu, J.: Evasion and obfuscation in speaker recognition surveillance and forensics. In: IEEE 2nd International Workshop on Biometrics and Forensics, Valletta, Malta (2014)
7. Singh, N., Khan, R.A., Shree, R.: Applications of speaker recognition. Procedia Eng. **38**, 3122–3126 (2012)
8. Li, L., et al.: CN-Celeb: multi-genre speaker recognition. Speech Commun. **137**, 77–91 (2022)
9. Kanervisto, A., Vestman, V., Hautamäki, V., Kinnunen, T.: Effects of gender information in text-independent and text-dependent speaker verification. In: IEEE International Conference on Acoustics, Speech and Signal Processing (ICASSP), New Orleans, LA, USA (2017)
10. Chowdhury, L., Zunair, H., Mohammed, N.: Robust deep speaker recognition: learning latent representation with joint angular margin loss. Appl. Sci. **10**(21), 1–17 (2020)

11. Paulose, S., Mathew, D., Thomas, A.: Performance evaluation of different modeling methods and classifiers with MFCC and IHC features for speaker recognition. Procedia Comput. Sci. **115**, 55–62 (2017)

12. Ayadi, M., Hassan, A., Abdelnaby, A., Elgendy, O.: Text-independent speaker identification using robust statistics estimation. Speech Commun. **92**, 52–63 (2017)

13. India, M., Safari, P., Hernando, J.: Self multi-head attention for speaker recognition. In: INTERSPEECH (2019)

14. Torfi, A., Dawson, J., Nasrabadi, N.:Text independent speaker verification using 3D convolutional neural network. In: 2018 IEEE International Conference on Multimedia and Expo (ICME), San Diego, CA, USA (2018)

15. Shon, S., Tang, H., Glass, J.: Frame-level speaker embeddings for text-independent speaker recognition and analysis of end-to-end model. In: 2018 IEEE Spoken Language Technology Workshop (SLT), Greece, Athens (2019)

16. Emre, S., Soufleris, P., Duan, Z., Heinzelman, W.: A deep neural network model for speaker identification. Appl. Sci. **11**(8), 1–18 (2021)

17. Ye, F., Yang, J.: Front-end speech enhancement for commercial speaker verification systems. Speech Commun. **99**, 101–113 (2018)

18. Liu, C., Yin, Y., Sun, Y., Ersoy, O.: Multi-scale ResNet and BiGRU automatic sleep staging based on attention mechanism. PloS One **17**, 1–20 (2022)

19. Kumar, T., Bhukya, R.: Mel spectrogram based automatic speaker verification using GMM-UBM. In: 2022 IEEE 9th Uttar Pradesh Section International Conference on Electrical, Electronics and Computer Engineering (UPCON), Prayagraj, India (2022)

20. Wang, J.C., Wang, C.Y., Chin, Y.H., Liu, Y.T., Chen, E.T., Chang, P.C: Spectral-temporal receptive fields and MFCC balanced feature extraction for robust speaker recognition. Multimedia Tools Appl. **76**, 4055–4068 (2017)

21. Gurbuz, S., Gowdy J., Tufekci, Z.: Speech spectrogram based model adaptation for speaker identification. In: Proceedings of the IEEE SoutheastCon 2000 'Preparing for The New Millennium', Nashville, TN, USA (2002)

22. Alam, J., Kinnunen, T., Kenny, P., Ouellet, P., O'Shaughnessy, D.: Multitaper MFCC and PLP features for speaker verification using i-vectors. Speech Commun. **55**(2), 237–251 (2013)

23. Hossan, A., Memon, S., Gregory, M.: A novel approach for MFCC feature extraction. In: 2010 4th International Conference on Signal Processing and Communication Systems, Gold Coast, QLD, Australia (2010)

24. Weng, Z., Li, L., Guo, D.: Speaker recognition using weighted dynamic MFCC based on GMM. In: 2010 International Conference on Anti-Counterfeiting, Security and Identification, Chengdu, China (2010)

25. Abdul, R., Setianingsih, C., Nasrun, M.: Speaker recognition for device controlling using MFCC and GMM algorithm. In: 2020 2nd International Conference on Electrical, Control and Instrumentation Engineering (ICECIE), Kuala Lumpur, Malaysia (2021)

26. Ajgou, R., Sbaa, S., Ghendir, S., Chamsa, A., Taleb A.: Robust remote speaker recognition system based on AR-MFCC features and efficient speech activity detection algorithm. In: 2014 11th International Symposium on Wireless Communications Systems (ISWCS), Barcelona, Spain (2014)

27. Sharma, D., Ali, I.: A modified MFCC feature extraction technique For robust speaker recognition. In: 2015 International Conference on Advances in Computing, Communications and Informatics (ICACCI), Kochi, India (2015)

28. Zhao, X., Wang, D.: Analyzing noise robustness of MFCC and GFCC features in speaker identification. In: 2013 IEEE International Conference on Acoustics, Speech and Signal Processing, Vancouver, BC, Canada (2013)
29. Li, Q., Huang, Y.: An auditory-based feature extraction algorithm for robust speaker identification under mismatched conditions. IEEE Trans. Audio Speech Lang. Process. **19**(6), 1791–1801 (2011)
30. Valero, X., Alias, F.: Gammatone cepstral coefficients: biologically inspired features for non-speech audio classification. IEEE Trans. Multimedia **14**(6), 1684–1689 (2012)
31. Ayoub, B., Jamal, K., Arsalane, Z.: Gammatone frequency Cepstral coefficients for speaker identification over VoIP networks. In: 2016 International Conference on Information Technology for Organizations Development (IT4OD), Fez, Morocco (2016)
32. Wang, H., Zhang, C.: The application of Gammatone frequency cepstral coefficients for forensic voice comparison under noisy conditions. Aust. J. Forensic Sci. **52**(5), 553–568 (2020)
33. Choudhary, H., Sadhya, D., Vinal, P.: Automatic speaker verification using gammatone frequency cepstral coefficients. In: 2021 8th International Conference on Signal Processing and Integrated Networks (SPIN), Noida, India (2021)
34. Farsiani, S., Izadkhah, H., Lotfi, S.: An optimum end-to-end text-independent speaker identification system using convolutional neural network. Comput. Electr. Eng. **100**, 107882 (2022)
35. Ashar, A., Shahid, M., Mushtaq, U.: Speaker identification using a hybrid CNN-MFCC approach. In: 2020 International Conference on Emerging Trends in Smart Technologies (ICETST), Karachi, Pakistan (2020)
36. Dwijayanti, S., Yunita, A., Yudho, B.: Speaker identification using a convolutional neural network (2022)
37. Soleymani, S., Dabouei, A., Mehdi, S., Kazemi, H., Dawson, J.: Prosodic-enhanced Siamese convolutional neural networks for cross-device text-independent speaker verification. In: 2018 IEEE 9th International Conference on Biometrics Theory, Applications and Systems (BTAS), Redondo Beach, CA, USA (2019)
38. Salvati, D., Drioli, C., Luca, G.: A late fusion deep neural network for robust speaker identification using raw waveforms and gammatone cepstral coefficients. Expert Syst. Appl. **222**, 119750 (2023)
39. Costantini, G., Cesarini, V., Brenna, E.: High-level CNN and machine learning methods for speaker recognition. Sensors **23**(7), 3461 (2023)
40. Bunrit, S., Inkian, T., Kerdprasop, N., Kerdprasop, K.: Text independent speaker identification using deep learning model of convolutional neural network. Int. J. Mach. Learn. Comput. **9**(2), 143–148 (2019)
41. Wondimu, L., Ramasamy, S., Worku, J.: Analyzing noise robustness of Cochleogram and Mel spectrogram features in deep learning based speaker recognition. Appl. Sci. **13**, 1–16 (2022)
42. Zhao, Z., et al.: A lighten CNN-LSTM model for speaker verification on embedded devices. Futur. Gener. Comput. Syst. **100**, 751–758 (2019)
43. Bader, M., Shahin, I., Ahmed, A., Werghi, N.: Hybrid CNN-LSTM speaker identification framework for evaluating the impact of face masks. In: 2022 International Conference on Electrical and Computing Technologies and Applications (ICECTA), Ras Al Khaimah, United Arab Emirates (2022)
44. Shekhar, H., Roy, P.: A CNN-BiLSTM based hybrid model for Indian language identification. Appl. Acoust. **182**, 108274 (2021)

45. Liu, Y.-H., Liu, X., Fan, W., Zhong, B., Du, J.-X.: Efficient audio-visual speaker recognition via deep heterogeneous feature fusion. In: Zhou, J., et al. (eds.) CCBR 2017. LNCS, vol. 10568, pp. 575–583. Springer, Cham (2017). https://doi.org/10.1007/978-3-319-69923-3_62

46. Nagrani, A., Chung, J.S., Zisserman, A.: VoxCeleb: a large-scale speaker identification dataset. In: INTERSPEECH (2018)

47. Kim, S.H., Park, Y.H.: Adaptive convolutional neural network for text-independent speaker recognition. In: INTERSPEECH (2021)

48. Ding, S., Chen, T., Gong, X., Zha, W., Wang, Z.: AutoSpeech: neural architecture search for speaker recognition. arXiv:2005.03215v2 [eess.AS], vol. 31 (2020)

State-of-the-Art Approaches to Word Sense Disambiguation: A Multilingual Investigation

Robbel Habtamu[iD] and Beakal Gizachew[(✉)][iD]

Addis Ababa Institute of Technology, Addis Ababa, Ethiopia
`beakal.gizachew@aait.edu.et`

Abstract. Word sense disambiguation (WSD) is determining the correct sense of an ambiguous word from its surrounding context. It's a fundamental problem in NLP that has wide-ranging effects on critical tasks like information retrieval, machine translation, and question-answering systems. The inherent complexity of language is like, word homonymy, polysemy, and other linguistic features, making it challenging to develop precise word sense disambiguation solutions. Our goal in this study is to find out more about the most recent developments in word sense disambiguation. Notable progress has been recently made in this field. However, dealing with languages with complex morphological structures, numerous dialects, or limited linguistic resources continues to be a challenging task for WSD. To the best of our knowledge, we present a comprehensive understanding of WSD across different languages including English, Chinese, Arabic, selected low-resource Indian languages, and Amharic. We look into various WSD techniques, including Knowledge-based, supervised, unsupervised, Deep Learning, and Hybrid approaches. The finding of this study shows that deep learning approaches and pre-trained models show promise in improving multilingual Word Sense Disambiguation, We also draw attention to the significant effort needed to address cross-lingual challenges in this field.

Keywords: Word Sense Disambiguation · Natural Language Processing · Application of WSD · State-of-the-art Approachs of WSD

1 Introduction

We are witnesses to the advancement of artificial intelligence (AI) in our daily routine and making life easier by making machines perform human functional activities in different sectors and allowing them to engage in autonomous learning and decision-making. Natural Language Processing (NLP) is a sub-field of AI that aims to bridge the gap between computers and human languages. In the field of natural language processing, word sense disambiguation (WSD) presents an important challenge that is intended to identify the proper sense of an ambiguous word within the sentence or context.

Word Sense Disambiguation is a challenging task in natural language processing, aiming to determine the proper sense of ambiguous words in a given

T. G. Debelee et al. (Eds.): PanAfriConAI 2023, CCIS 2068, pp. 176–202, 2024.
https://doi.org/10.1007/978-3-031-57624-9_10

context. For example, consider the English word *Bass* in the following sentences:**Sentence-1: I can hear bass sounds. Sentence-2: They like grilled bass.** Within those two sentences, the word "**Bass**" has two different meanings, referring to low-frequency tones and a type of fish, respectively [75]. WSD is, an AI-complete problem [74]. Its accuracy significantly impacts various natural language processing applications, such as machine translation [103], information retrieval [115], and sentiment analysis [49].

This study offers a comprehensive analysis of word sense disambiguation approaches utilized in different languages, especially by paying attention to different language-specific challenges faced during previous work. Additionally, we explore cross-lingual WSD and We discuss possible directions for future research.

2   Historical Development of WSD

Research on word sense disambiguation has a long and dynamic history. The renowned machine translation memo written by Warren Weaver in the early 1940 s is where this field of study got its start [107]. Weaver's study demonstrates a good understanding of the importance of context and the underlying statistical character of WSD difficulties. In addition to establishing the foundational ideas that remain in use today, he also promoted the beginning of statistical semantic investigations as a necessary first step [83].

In 1957, Masterman introduced a groundbreaking concept, as documented in [84]. He proposed a method that relied on the headings of categories in Roget's International Thesaurus to determine a word's precise meaning. This novel strategy was an important development in the early stages of Word sense disambiguation. However, significant progress in WSD research wasn't made until the 1980 s [81], when large lexical resources and corpora were available. These sources provide researchers with much information and resources, allowing them to advance the WSD challenge significantly. Lesk [59] developed a system in 1986 that was based on the overlaps between dictionary definitions of the terms in a sentence. In the 1990 s, a large number of scientists began using both machine-readable and human-coded dictionaries simultaneously [109]. Another popular method used for addressing WSD is the use of dictionary-based corpora, which was first detailed by Brown et al. [108]. These corpora contain content that has had word sense annotations from lexicons or dictionaries.

The development of knowledge-based techniques and rule-based systems made use of lexical resources. In the 1990 s, there were significant developments in machine learning and computational linguistics, which created new opportunities for data-driven methods of word sense disambiguation. Large corpora and annotated datasets are readily available in the contemporary context, and statistical and machine-learning techniques have shown remarkable potential. Kilgarriff [55] states that Senseval was the first open, community-based evaluation of Word Sense Disambiguation programs, making it a significant milestone. Throughout the summer of 1998, there were tasks in Italian, French, and English for this historic event.

Researchers have made significant progress in improving word sense disambiguation using deep learning and neural network techniques. Contextual word embeddings like Context2Vec and ELMo have shown promising performance in several WSD tasks. BERT is a transformer-based model that provides a significant improvement in WSD accuracy. Other contextualized models were developed, including RoBERTa and GPT. However, WSD remains a demanding task with many open problems, including polysemy and homonymy. Linguistic complexities are not easily determined by contextually appropriate meaning, and most current WSD models cannot be used. Aspects regarding contextual variation, dynamic changes in the meaning of words, and demands for cross-domain proficiency compose a complex picture of WSD (Fig. 1).

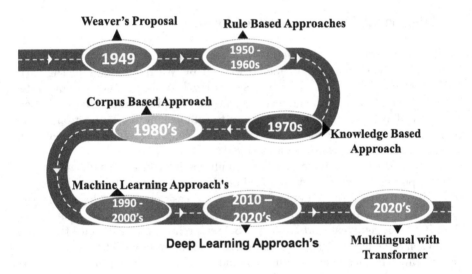

Fig. 1. WSD Approaches Development

2.1 Fundamental Tasks in WSD

The following list of essential steps must be taken to complete the process of word sense disambiguation.

Word Sense Identification: The first crucial step and central focus of WSD is determining a word's sense in a particular context. Unlike many other classification problems, word senses are complicated and challenging to discretize into a small, discrete set of entries, each expressing a distinct meaning [94].

Use of External Knowledge Resources: Word sense disambiguation heavily depends on external resources. One of the most important resources is using sense inventories, like WordNet and BabelNet, which systematically organize words into synsets. To effectively traverse the complex network of language, sensee inventories are essential in offering an accurate representation of word senses [75]. Both structured and unstructured sources can be used as knowledge sources. Ontologies, thesauri, and machine-readable dictionaries (MRDs) are a few examples of structured sources. On the other hand, corpora and collocation resources are unstructured sources.

Sense Annotation: Context words are manually labeled with the most suitable meanings. This entails a series of preliminary tasks such as chunking, parsing, normalization, lemmatization, tokenization and part-of-speech tagging. Then global features, syntactic features, semantic features, and local features are considered.**Choosing Appropriate Approach:**Different approaches include supervised approach, knowledge-based approach, unsupervised approach and deep learning approach in word sense disambiguation. Thus it is important to choose an appropriate approach according to our data and specific task.**Evaluation Metrics:** In word sense disambiguation evaluation measures like accuracy, precision, recall F1 score, etc., are used for assessing the quality and performance of WSD algorithms. Nevertheless, the lack of gold-standard annotated data makes it difficult to evaluate and compare results in word sense disambiguation (Fig. 2).

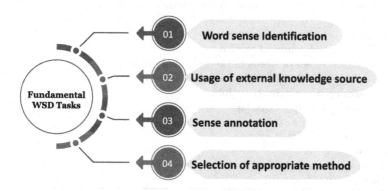

Fig. 2. Task Summary in WSD

3   Approachs of WSD

This section explores the various methods employed in Word sense disambiguation techniques generally fall into three main categories: knowledge-based, machine learning, and deep learning approaches.

3.1   Knowledge Based Approach

The knowledge-based approach to Word Sense Disambiguation relies on external knowledge resources such as WordNet or other lexical databases. These methods use lexical knowledge bases such as dictionaries and thesauri and extract knowledge from word definitions and relations among words and senses [78]. These resources organize words into synsets, providing hierarchical relationships and semantic information. Basic knowledge-based techniques include Lesk algorism [59] which considers overlapping word senses in the surrounding context, Measure of semantic similarity computed over semantic network, Heuristic method [66], and Automatically or semi-automatically acquired [18]. The Lesk Algorithm, WordNet-based Approaches, and Ontology-based Approaches are three popular knowledge-based methods that we will now describe in detail.

Lesk Algorithm: Michael Lesk first presented the Lesk Algorithm as a traditional method in 1986. The idea of **"word over-lap"** is employed to determine the appropriate meaning of an ambiguous word in a particular context. The Lesk algorithm examines word pairs to determine the meaning of a phrase. It compares the definitions of surrounding words against possible senses of the target word. This allows it to select the best sense in context.

Algorithm 1. SIMPLIFIED LESK Algorithm

---

function SIMPLIFIED _ LESK(word, sentence)
    best_sense ← most frequent sense for word
    max_overlap ← 0
    context ← set of words in sentence
    for each sense in senses of word **do**
        signature ← set of words in the gloss and examples of sense
        overlap ← COMPUTEOVERLAP(signature, context)
        if overlap > max_overlap **then**
            max_overlap ← overlap
            best_sense ← sense
        end if
    end for
    return best_sense
end function

Because the Lesk Algorithm is computationally efficient and conceptually simple, it is appropriate for low-resource languages. However, the algorithm does not take word sense polysemy into account and also struggles when there is no sufficient data (Table 1).

3.2 Supervised Approach

Using a corpus of words tagged with their sense in context, the supervised method of WSD trains a classifier to tag words in new texts. To predict the accurate sense for possible instances, these models take patterns and features from the labeled data. Features in the models could include grammatical information, part-of-speech tags, or word context. The supervised method has the advantage

Table 1. Base line paper in knowledge base approach with their evaluation

Model	Evaluation (F1-score for each gold standard datasets)						
	Senseval-2	Senseval-3	Senseval-2007	Senseval-2013	Senseval-2015	All	Year
Lesk_ext	50.6	44.5	32.0	53.6	51.0	48.7	2003 [10]
Lesk_ext+emb	63.0	63.7	56.7	66.2	64.6	63.7	2014 [12]
UKB	68.8	66.1	53.0	68.8	70.3	57.5	2014 [3]
UKB_gloss	64.2	54.8	40.0	64.5	64.5	57.5	2014 [3]
WN 1st sense baseline	66.8	66.2	55.2	63.0	67.8	65.2	2017 [87]
Babelfy	67.0	63.5	51.6	66.4	65.5	70.3	2014 [71]
WSD-TM	69.0	66.9	55.6	69.6	65.3	66.9	2018 [22]
kEF	56.9	72.3	69.6	66.1	68.4	68.0	2020 [104]
SCSMM	68.9	67.6	57.1	63.5	69.5	66.7	2022 [6]

of using machine learning techniques to effectively disambiguate and capture complex relationships, but it requires labeled training data, which can be costly and time-consuming to produce [105]. Algorithms like XGBoost, AdaBoost, Neural Networks, Naive Bayes, Support Vector Machines, and Decision Trees are frequently employed in supervised learning [14]. We addressed a few of these techniques below, summarising their benefits and drawbacks concerning the word sense disambiguation task.

Naive Bayes: This approach is highly effective and computationally efficient on small datasets. Its adaptability to discrete and continuous data makes it suitable for various linguistic contexts. However, due to its simplistic assumption of feature independence, it may perform poorly with strongly correlated features and struggle to recognize complex language relationships. Below is the Naive Bayes Assumption algorithm.

$$P(s|c) = \frac{P(c|s) \cdot P(s)}{P(c)} \qquad (1)$$

Where,

$P(s|c)$: Probability of sense s given context c

$P(c|s)$: Likelihood of context **c** given sense **s**

$P(s)$: Prior probability of sense s

$P(c)$: Probability of context c

Support Vector Machines (SVM): Both linear and non-linear separation boundaries are well managed by SVM. Word sense disambiguation [34,58], parsing [25], and text summerization [50] are some of the NLP tasks for which SVM can be applied. Compared to other supervised methods, it has been shown that SVM yields the best results in WSD [58]. Nevertheless, SVM needs large annotated datasets for training and can be computationally expensive.

Decision Trees: According to Jumi Sarmah [92], a decision tree resembles a flow chart with internal nodes representing tests, branches representing test results, and leaves representing sensory levels. This method provides a straightforward and comprehensible tool for word sense disambiguation. However, using deep trees can increase the risk of overfitting, potentially leading to poor generalization performance.

Neural Networks: Neural network models provide exceptional performance in several natural language processing problems by deriving intricate representations from the data. Furthermore, as neural networks are multilingual, they can adapt to different linguistic contexts [38] (Table 2).

Table 2. Baseline paper for Supervised approach with their evaluation

Model	Evaluation (F1-score for each gold standard datasets)					
	Senseval-2	Senseval-3	Senseval-2007	Senseval-2013	Senseval-2015	Year
ELMo	71.6	69.6	62.2	66.2	71.3	2018 [82]
Bi-LSTM_att+LEX+POS	66.9	69.1	64.8	71.5	72.0	2017 [87]
BiLSTM_att+LEX	72.0	69.4	63.7	66.4	72.4	2017 [86]
GAS	72.0	70.0	-	66.7	71.6	2018 [65]
BERT	73.8	71.6	63.3	69.2	74.4	2019 [43]
EWISER	78.9	78.4	71.0	78.9	79.3	2020 [13]
SemCor_WNGC+hypernyms	79.7	77.8	73.4	78.7	82.6	2019 [102]
EWISER+WNGC	80.8	79.0	75.2	80.7	81.8	2020 [13]
ESR	81.3	79.9	77.0	81.5	84.1	2021 [96]
ESR+WNGC	82.5	80.2	78.5	82.3	85.3	2021 [96]
ConSeC+WNGC	82.7	81.0	78.5	85.2	87.5	2021 [11]

3.3   Unsupervised Approach

Unsupervised methods offer a way to bypass the knowledge acquisition bottleneck [36]. This method operates without the need for labeled data and instead utilizes clustering or topic modeling methods [23]. The primary goal is to group words with similar contexts, assuming they share similar senses. Commonly used unsupervised methods include context clustering, word clustering, and co-occurrence graphs [5]. This method can detect potential senses without explicit annotations but encounters challenges in selecting the best sense for each group. some [67] use info from the inventory for mapping the corpus data to the gold standard and others use some level of explicit knowledge [111].

3.4   Semi-supervised Approach

To enhance WSD performance, the semi-supervised approach combines labeled and unlabeled data. It utilizes a bigger amount of unlabeled data. *What if you don't have the data availability that supervised and dictionary-based techniques require?*; well, employing semisupervised methodology known as **bootstrapping** is best solution. A bootstrapping model was presented by Yarowsky in

1995 [112]. It starts with a few sense-labeled examples and gradually assigns appropriate senses to cases that are not yet labeled. The fundamental aim of semi-supervised learning is to use a small number of human-labelled instances as seeds to automatically categorize the unlabeled examples.

3.5 Hybrid Approach

To improve disambiguation accuracy and robustness, the hybrid approach in WSD combines several techniques, such as combining multiple models, merging outputs from different algorithms, and integrating knowledge-based resources like WordNet into supervised models [80,101]. Hybrid methods often leverage the strengths of both knowledge-based and corpus-based approaches, aiming to compensate for each other's limitations and provide more robust disambiguation.

3.6 Deep Learning Approach

Deep learning methods have significantly improved in the last several years in several natural language processing applications, such as word sense disambiguation. Neural networks and transformers are two examples of deep learning models that have demonstrated remarkable powers in identifying complex patterns and semantic links in textual data, which makes them a viable solution for word sense disambiguation problems.

Recurrent neural networks (RNNs) are one of the prominent deep learning techniques for WSD. This method involves training a neural network using a sizable dataset that includes instances of word usage in various circumstances. Words with similar meanings are closer to one another in a continuous vector space, which is how the network learns to represent words and their situations. The semantic content of words is encoded by this vector representation, sometimes called word embeddings. By automatically learning feature representations and utilizing contextual data, deep learning techniques have shown state-of-the-art performance in a variety of NLP tasks [90,110], including WSD. CNNs, RNNs, and transformer models such as BERT and GPT show promise in capturing complex word relations and representations [89].

Because pre-trained models make it possible to extract rich language representations from enormous amounts of unlabeled input, they have revolutionized natural language processing. Pre-trained models provide the following benefits when it comes to word sense disambiguation tasks: they minimize the need for handmade features, learn features, acquire semantic similarity, support multilingual data, and capture contextual information.

Attention mechanisms and other context-aware strategies are essential for optimizing deep learning models. With the attention mechanism, deep learning in WSD can be greatly enhanced by allowing models to assign varying importance to different words. Attention extends beyond individual words, allowing models to consider relationships across entire sentences or documents. Notably, many researchers now incorporate attention mechanisms into their methods to improve

WSD accuracy and capture intricate semantic relationships across diverse textual contexts (Table 3, 4 and 5).

Table 3. State of the art Deep learning methods for word sense disambiguation.

Model	F1-score			Accuracy		
	Senseval-2	Senseval-3	All	general-purpose	domain-specific	year
BLSTM	66.9	73.4	–	–	–	2019 [20]
FastText	–	–	53.7	56.2	50.6	2020 [17]
Bert-base	–	–	75.3	73.3	77.9	2020 [17]
Transformer	–	–	77.8	75.2	81.0	2020

Table 4. Comparisons of WSD approaches

Criteria	Approaches			
	knowledge-based	Supervised	Unsupervised	Deep learning
Data Needed	annotated	Labeled data	Corpus	mixed
Efficiency	High	Moderate	High	Moderate
Scalability	Low	Low	High	High
Language Adaptability	Limited	Limited	High	High

Note: The "**Mixed**" category in the "**Data Needed**" column indicates a combination of annotated and unlabeled data. **Scalability** is labeled based on the need for annotated or labeled data during model training. Language Adaptability refers to the adaptability of the approach to different languages.

4   WSD Across Different Languages

An overview of WSD in English, Chinese, Arabic, Indian languages, and Amharic is discussed in this section. We have discussed the general techniques and methods utilized for WSD in many languages and have emphasized the main barriers and difficulties unique to each language.

4.1   WSD in English Language

In the English language, word sense disambiguation is a well-researched NLP task. Much of the history of WSD has been determined by the availability of manually created lexical resources in English, including SemCor [22,70] and **Word-Net** [35,69]. Then the introduction of BabelNet [76] was a massive multilingual semantic network, created by automatically integrating WordNet, Wikipedia, and other resources [63].

Table 5. Summary of WSD Approaches

Approach	Advantages	Disadvantages
Knowledge-based	• Can provide explicit sense definitions and relationships [78] • Interpretable and explainable	• Limited data coverage. • Difficulty in handling new words • Can't capture contextual features
Supervised	• High accuracy with sufficient data • Can capture complex context • Flexibility to use various features	• Requires large annotated data • Difficulty in handling new words
Unsupervised	• Doesn't require labeled data [23] • Ability to discover sense clusters • Can handle new word senses	• Less accurate compared to supervised methods. • Limited Interpretability and explainability.
Semi-Supervised	• leverage large amount of unlabeled data. • Reduces the need for manual annotation.	• Performance is highly dependent on the quality of data.
Deep Learning	• Ability to automatically learn complex representations [89]. • Can handle multiple languages and contexts effectively.	• Requires large amounts of data for effective training. • Computationally expensive for training and inference.

In 1995, Yarowsky [44] achieved over 95% accuracy in Word Sense Disambiguation using a semi-supervised approach for 12 words. Bootstrapping was employed to train a high-precision word sense disambiguation classifier.

According to Kilgarriff and Palmer (2000) [2], the Senseval-1 evaluator achieved a 77% accuracy rate in the English lexical sample task in 1997, while human performance, determined by inter tagger agreement, reached 80%.

Stevenson and Wilks employed Part-of-Speech data on all word WSD in 2001 and got an accuracy rating of 94.7% [97]. Due to the utilization of WordNet, Senseval-2 produced a lower score (Edmonds and Cotton 2001) [80], prompting the necessity for further development in Senseval-3. This development ultimately resulted in its top systems performing at human levels on the English lexical sample test [2].

State-of-the-art English WSD methods utilize deep learning and transformer-based architectures, like BERT (Bidirectional Encoder Representations Transformer) and its variants, which have shown exceptional performance in various NLP tasks, including WSD [47]. BERT uses extensive pretraining on a large corpus to learn contextualized word representations. These pre-trained models can be fine-tuned with specific WSD data to enhance disambiguation accuracy.

Additionally, **Ensemble methods** [52], which combine approaches, have improved English WSD performance. Recent research has also explored the inte-

Table 6. Summary of paper in English language WSD

Paper	Technique	Dataset	Language
Waver (1940) [107]	Statistical Semantics	Own	English
Y. Wilks (1975) [108]	Semantic Preference	Own	English
Lesk (1986) [59]	Knowledge based	OALD	English
Miller et al. (1990) [70]	Stastical Methods	Corpora	English
Brown et al. (1991) [19]	Statical method	Corpora	English
D. Yarowsky (1995) [111]	unsupervised learning	Own	English
B. A. Onyshkevych (1997) [79]	Knowledge based	Coprpora	English
M. Stevenson, Wilks (2001) [97]	Knowledge based	Corpora	English
C. Stokoe et al. (2003) [98]	Statistical Method	Own	English
C. Fellbaum, G. Miller (2003) [35]	Knowledge based	Wordnet	English
Kilgarriff et al. (2003) [30]	Supervised approach	Senseval	English
D. Vickreyet al. (2005) [103]	Context Aware	Corpora	English
Agirre et al. (2006) [2]	unsupervised approach	Own	English
Kolte et al. (2008) [56]	Unsupervised approach	Wordnet	English
Z. Zhong, H.T. Ng (2012) [115]	Language Modeling	Own	English
R. Navigli et al. (2012) [76]	Automatic construction	Wordnet	Multilingual
GR Haffari, A Sarkar (2012) [44]	Semi-supervised approach	Own	English
R. Chain et al. (2013) [23]	unsupervised methods	Own	English
A R. Pal et al. (2015) [80]	Hybrid approach	Own	English
D.S. Chaplot et al. (2018) [22]	Knowledge based	Wordnet	English
J. Devlin et al. (2018) [29]	BERT	Corpus	English
F. Luo et al. (2018) [33]	Co-attantion mechanism	Own	English
L Huanget al. (2019) [48]	BERT based models	Own	English
C. Hadiwinoto et al. (2019) [43]	Pretrained models	Multiple	English
Y. Liu et al. (2019) [60]	RoBERTa	Corpus	English
Y. Luan et al. (2020) [63]	Novel	Babelnet	Multilingual
Z. Lan et al. (2019) [57]	ALBERT	Corpus	English
M Saeidi et al. (2023) [91]	Wikification	Wikipedia	English
H. Kanget al. (2023) [52]	Pre-trained model	XL-WSD	Multilingual
D. Loureiro (2023) [61]	Deep learning	Wordnet	English

gration of external knowledge resources, such as WordNet or BabelNet, to enrich the semantic information available for English WSD [61,91].

Even with all of the study and implementation of Word Sense Disambiguation in English, important issues still need to be addressed. Intending to identify tiny variations in fine-grained senses, researchers concentrate on improving sense granularity. Improving contextual information handling is essential to successfully adjusting WSD models to various situations. In the following table, we have summarised some papers that have focused on word sense disambiguation in the English language (Table 6).

4.2   WSD in Chinese Language

Whatever the language, there is an ambiguous word in it, For example, the Chinese word "zu" has two common meanings, which are "si wang" and "shi bing" [100]. Word sense disambiguation in Chinese has special challenges due to the language's grammatical uniqueness. The inherent ambiguity resulting from the nature of Chinese characters is one significant difficulty. Furthermore, Chinese language lacks clear word borders, adding complexity that makes strong word segmentation necessary to accurately identify individual words for disambiguation. Furthermore, because a word's meaning can change depending on its environment, contextual changes and syntactic flexibility in Chinese phrases present difficulties.

Researchers have explored various techniques in Chinese WSD, ranging from neural networks to domain adaptation, from word embeddings to cross-language transfer learning. One type of Chinese knowledge resource is typically used in the conventional graph-based Chinese Word sense disambiguation method, which is extremely affected by the knowledge bottleneck issue [113]. Compared with knowledge resources in Chinese, those in English are more mature and abundant. To solve this problem, Wenpeng Lu. et al. [62] proposes a graph-based Chinese WSD method with multi-knowledge integration. Their promising results were obtained with comprehensive SemEval dataset experimentation on a graph model that combines multiple Chinese and English knowledge resources through word meaning mapping.

Similar to English WSD, Chinese WSD benefits from the use of word embeddings such as Word2Vec or GloVe. These embeddings capture semantic relationships between words and can be used to enhance the representation of Chinese words in a continuous vector space. Due to poor coverage of Chinese by BabelNet, HowNet [46] has been used in a parallel line of study as both a sense inventory and lexical knowledge base for Chinese WSD. HowNet is a widely used Chinese lexical knowledge base. Zhang et al. [114] introduced a novel approach in 2022, integrating monolingual contextual data from a neural language model, bilingual information from machine translation, and sense translation data from HowNet. This approach is a departure from traditional HowNet-based WSD methods. Yiming Cui et al. [26] proposed Pre-Training With Whole Word Masking for Chinese BERT, and their model, MacBERT, has shown state-of-the-art performance in various NLP tasks.

Recent supervised neural WSD algorithms improve performance by making use of a lexical knowledge base, such as by integrating definitions [64]. In 2019, L. Huang et al. build context-gloss pairs and suggest three BERT-based models for WSD. They fine-tune the pre-trained BERT model to produce new, state-of-the-art outcomes on the WSD problem [48]. The study claims that graph-based models are effective in Chinese WSD because they generate graphs with nodes representing words or senses and edges representing semantic links.

Additionally, Chinese WSD studies, using pre-trained models have gained more attention. The work by Christian Hadiwinoto et al. [43] is significant since it proposes their work titled: "Improved Word Sense Disambiguation Using Pre-

Trained Contextualised Word Representations." Their technique uses a neural sentence encoder, which is essential for word disambiguation because it can capture context very well. This progress shows a great deal of potential and adds to the body of knowledge in the field of word sense disambiguation.

4.3   WSD in Arabic Language

Since every language has its idioms, grammatical structures, and word usage patterns, it is difficult to directly apply WSD techniques across languages. Word sense disambiguation is a global problem that crosses linguistic boundaries, with Arabic and other Semitic languages facing especially difficult problems [4].

The lack of diacritics in many digital texts is a major cause of word ambiguity in Arabic, allowing the same word to appear in many senses [33]. It allows identical words to take on various meanings; without diacritics, they will be the same word. Furthermore, some Arabic prefixes are closely connected to Arabic words and can be regarded as words in the majority of Latin languages, which is to blame for the agglutinative nature of the Arabic language [31].

Arabic words have contextual variations according to diacritical markings, part-of-speech (POS) characteristics, and the context in which they occur (Habash, 2007 [41]). Furthermore, Zitouni in 2014 [116] emphasised that the lack of vowel markers in Arabic script adds considerably to the language's intrinsic morphological ambiguity. Remarkably, a 2002 study [27] by Debili and colleagues found that Arabic texts that have vocalization marks (i.e., marks that indicate vowel sounds) typically had less ambiguous words (43%) than texts that do not have vocalization markers (72%). Using thematic words from a specific context, Sakhr researcher Achraf Chalabi created a new word sense disambiguation algorithm in 1998 that has been implemented in the system's Arabic-to-English computer-aided translation [21].

Arabic language still suffers from the lack of linguistic resources [31] such as lexicons, thesauri, tagged corpora [9], and standardized test collections [15]. Furthermore, due to this language's extensive use of inflection, complex morphology in Arabic has a significant impact on IR systems [16]. The richness of Arabic texts and the **free word order** phenomenon are the main issues with statistical methods for Arabic word sense disambiguation. The most prominent supervised approach for Arabic WSD is the work by Habash et al. [42]. Their achievement of statistically meaningful outcomes on well-known benchmark datasets, such the Arabic Treebank, sets them apart from the others.

The established success of English has made contextualized embeddings a desired option for Arabic in recent times. In this field, numerous models of contextual embedding have been developed, including BERT [29], RoBERTa [60], ALBERT [57], hULMonA [32], and AraBERT [7], have been created in this area.

A recent study of Arabic word sense disambiguation highlights some key unresolved challenges that need more research. Two important areas are domain-specific Arabic word sense disambiguation and multi- and cross-lingual disambiguation for Arabic. These topics showcase both current difficulties and chances to advance Arabic word sense disambiguation through further examination.

Distinctive qualities like missing diacritical marks, multifaceted morphology, and contextual fluctuations make the Arabic language ambiguous. Arabic WSD faces challenges from limited resources. Researchers tackle this by making custom lexicons. So comparing approaches grows more complex with differences in lexicons, data, and goals. In summary, researchers widely prefer pre-trained models for Arabic WSD because fine-tuning them improves performance.

4.4   WSD in Indian Languages

India, which is known for its linguistic diversity, is home to 122 major languages and 1599 other languages, according to M. Priya et al. [85]. Approximately 70% of people speak Indo-Aryan languages, and 19% speak Dravidian languages, which include Bengali, Marathi, Telugu, Tamil, Gujarati, Kannada, and Malayalam. These languages are distinguished by their morphological richness and agglutinative structure. Word Sense Disambiguation has been used more frequently in English and other European languages than in Indian languages, according to Alon et al., [81]. The large variety of morphological inflections present in Indian languages, together with the lack of machine-readable dictionaries, word sense inventories, and other knowledge resources required for WSD algorithms in these languages, are the reasons for the suboptimal efficacy of WSD methods.

One of the morphologically rich languages in India is the **Marathi** language [37]. As per the research conducted by Ujwalla Gawande et al., [37], a novel approach to WSD for the Marathi language has been proposed using machine learning techniques. Their experimental results demonstrate notable performance improvements in this field.

The other language is **Manipuri**, which is distinct from other Indian languages in terms of both syntactic and semantic features and is spoken in a particular geographical area [81]. In a pioneering work by Richard Singh and K. Ghosh [95], a novel architecture was proposed in 2013 specifically for the Manipuri Language. The system operates in two distinct phases: the training phase and the testing phase, effectively performing Word Sense Disambiguation. The model performance was hindered by the lack of relevant morphological features.

Malayalam is a Dravidian language, primarily spoken in the southern Indian state of Kerala. It is one of the 22 official languages of India, and it is used by around 36 million people. Rosna P. Haroon in 2010 proposed Malayalam WSD using a knowledge-based approach and suggested that the approach used will result in poor accuracy due to the scarcity of corpus language like Malayalam. In 2016, S. Gopal [40] proposed a Supervised Malayalam word sense disambiguation system using the Naive Bayes classifier. Because the quality corpus was used, the accuracy was 90%.

Another Indian language with a rich morphological system is **Punjabi** [81]. In 2020, VP Singh et al. [94] recommended using deep learning approaches for word sense disambiguation in the Punjabi language. Their investigation demonstrates the effectiveness of these methods for correctly distinguishing between word senses in the context of Punjabi.

Challenges in WSD for low-resource Indian languages include complex morphology and limited resources. Positive progress is seen with deep learning models pre-trained on large datasets, offering effective solutions. Despite linguistic diversity and resource constraints, innovative techniques are driving notable advancements, especially in languages like Marathi, Manipuri, Malayalam, and Punjabi.

4.5   WSD in Amharic Language

Amharic, the second-most spoken Semitic language globally with approximately 22 million speakers in Ethiopia, serves as an official language of the Ethiopian federal government. Utilizing the distinctive **Ge'ez** script, an Abugida writing system with a history spanning over 2000 years, Amharic employs 33 consonant letters and 7 vowel letters, enabling the creation of over 300 distinct syllables. The language's intricate morphology, marked by a multitude of prefixes, suffixes, and infixes leading to varied word forms, poses a considerable challenge for word sense disambiguation in linguistic analysis.

WSD for the Amharic language was attempted by Kassie [53] in 2009 utilizing semantic vector analysis. He used a total of 865 words from Ethiopian Amharic-written legal statutes. As a result, their typical precision and recall were 58% and 82%, respectively. Then, in 2010 Amharic WSD study was carried out by Solomon [68] utilizing a corpus-based, supervised machine learning method and got 83.5% best accuracy for word level. In 2011, Assemu [8] introduced an unsupervised machine learning based corpus-driven approach for Amharic WSD, outperforming supervised methods in accuracy. Hagere [45] enhanced Solomon's supervised machine learning approach by employing ensemble classifiers like Adaboost and Bagging. Wassie's work [106] in 2014 introduced a semi-supervised WSD model for Amharic words, incorporating instance similarity-based clustering, unsupervised machine learning techniques, and supervised machine learning with unlabeled data.

Mulugeta [73] in 2019 used the Amharic WordNet hierarchy as a knowledge source to create an Amharic WSD system. They used an enhanced semantic space method that took context-to-gloss overlap into account. Mulugeta aimed to encompass all open-class words in the development of WordNet, a departure from earlier research that predominantly focused on verbs.

In 2021, Senay et al. [28] introduced an Amharic Word Sense Disambiguation system employing deep learning methods. They trained three distinct deep learning models (LSTM, CNN, and Bi-LSTM) on a dataset that encompassed 159 ambiguous words, 1214 synsets, and 2164 sentences. The experimental results demonstrated model accuracies of 94%, 95%, and 96% for LSTM, CNN, and Bi-LSTM, respectively. In 2022, NEIMA [72] conducted an Amharic WSD study using AmRoBERTa, a transformer-based contextual embedding technique,

Table 7. Summary of WSD Papers in Different Languages

Paper	Year	Technique	Dataset	Language
Zhi-Zhuo & He-Yan [113]	2012	Graph	Own	Chinese
XR Sun et al. [100]	2017	LSTM	Own	Chinese
Wenpeng Lu. et al. [62]	2019	Graph	BabelNet	Chinese
L. Huang et al. [48]	2019	BERT	Own	Chinese
Yiming Cui et al. [26]	2019	BERT	Own	Chinese
C. Hadiwinoto et al. [43]	2019	Pre-trained	OntoNotes	Chinese
B Hou et al. [46]	2020	Unsupervised	HowNet	Chinese
Y. Cui et al. [26]	2021	Pretrained	Own	Chinese
X. Zhang et al. [114]	2022	Novel	HowNet	Chinese
I. Bounhas et al. [15]	2011	Knowledge	ArabOnto	Arabic
Debili et al. [27]	2002	Knowledge	Corpus	Arabic
Habash et al. [41]	2007	Knowledge	Corpus	Arabic
Habash et al. [42]	2013	Knowledge	Own	Arabic
S. Elmougy et al. [33]	2008	Naive Bayes	Own	Arabic
Kharate & Patil [54]	2021	Knowledge	WordNet	Marathi
U. Gawande et al. [37]	2023	Novel	WordNet	Marathi
R. Singh & K. Ghosh [95]	2013	Decision Tree	Own	Manipuri
S. Gopal [40]	2016	Naive Bayes	Corpora	Malayalam
J. Sarmah & S.K. Sarma [92]	2016	Decision tree	Own	Assamese
VP Singh et al. [94]	2020	Deep learning	Corpus	Punjabi
Kassie [53]	2009	Knowledge	Own	Amharic
Solomon [68]	2010	Supervised	Own	Amharic
Assemu [8]	2011	Unsupervised	Own	Amharic
Hagerie [45]	2013	Ensemble	Own	Amharic
Wassie [106]	2014	Semi-supervised	Own	Amharic
Mulugeta [73]	2019	Deep learning	Own	Amharic
Senay et al. [28]	2021	Deep learning	Own	Amharic
NEIMA [72]	2022	Deep learning	Own	Amharic

demonstrating the potential for precise WSD in Amharic through advanced language models and transfer learning. **Broadly speaking**, this is the most widely used approach to Word Sense. Descriptive is a model approach based on knowledge and training. Knowledge approaches rely on vocabulary and inventories (Table 7).

5   WSD in NLP Applications

Word sense disambiguation plays a crucial role across various applications, including question answering, information retrieval, machine translation, and sentiment analysis. It enables precise language understanding by choosing the most fitting word meanings, thereby enhancing translation quality, decoding questions accurately, categorizing emotions correctly and providing suitable responses. Now, we will briefly examine some practical NLP applications that have benefited from word sense disambiguation.

1. **Machine Translation (MT)**
   Machine Translation benefits from WSD, as it helps improve translation quality by selecting the most appropriate translations for ambiguous words, such as in neural machine translation (NMT) systems. Research by Vickrey et al. [103] demonstrates its effectiveness in refining translation choices. For example, Quang-Phuoc Nguyen et al., [77] found that a Korean WSD system significantly enhances translation quality metrics like BLEU, TER, and DLRATIO in NMT. Incorporating word senses into machine translation holds great promise for enhancing the accuracy and quality of translations [88].

2. **Information Extraction(IE)**
   By extracting specified information like entities, relations, and properties, Information Extraction (IE), a branch of NLP, organizes unstructured text data. WSD, which clarifies word meaning ambiguities, is a fundamental objective of Information Extraction [24]. When a word has multiple possible meanings, ambiguity develops, which makes proper information extraction difficult.

3. **Information Retrieval(IR)**
   Information Retrieval is vital in our daily lives, used for tasks such as email communication, web searching, and database access.IR enhancements optimize data filing and retrieval, but word sense ambiguity, notably in web searches, compromises precision [98]. Research by M.A. Abderrahim et al. [1] indicates that WSD can improve IR performance.

4. **Sentiment Analysis**
   Sentiment analysis, an element of NLP, seeks to detect emotions and opinions in text, classifying them as positive, negative, or neutral [51]. Integrating WSD into sentiment analysis significantly enhances accuracy, surpassing systems without it [51,99].

5. **Speech Processing**
   WSD is crucial in speech processing within NLP, where spoken language is converted into text, as it helps resolve ambiguity and allows for correct linguistic analysis [93].

6. **Question Answering(QA)**
   In the field of NLP, Question Answering systems [39], which aim to comprehend and respond to human queries in natural language, greatly benefit from WSD. WSD plays a pivotal role in QA by ensuring the accurate interpretation of word meanings during the question analysis and answer retrieval phases.

6   Cross-Lingual and Multilingual WSD

Cross-lingual and multilingual word sense disambiguation addresses the challenges of understanding and disambiguating the meaning of words in multiple languages.

Cross-lingual word Sense Disambiguation strives to employ resources from one language to elucidate words in another. This approach depends on language-specific assets like cross-lingual embeddings, bilingual dictionaries, and parallel corpora. However, difficulties emerge due to divergences in language structures, changing word meanings, finite linguistic resources, and the lack of direct translations. Earlier methodologies encompass bilingual dictionaries, parallel corpora, multilingual embeddings, knowledge-grounded techniques, and machine translation. Recent progressions in Cross-Lingual WSD incorporate tactics for domain adaptation, unsupervised methodologies, and transfer learning procedures.

Multilingual Word Sense Disambiguation means resolving word meanings in different linguistic contexts all at once. To boost its efficiency, the use of Word-Net and pre-trained multilingual models is critical. Its scalability is possible due to its unified structure for various languages. However, the challenges that have to be dealt with include managing low-resource languages and coordinated sense distinctions. Prior methods composed techniques like adapting to domains, unsupervised techniques, and leveraging pre-trained multilingual models. Current developments in Multilingual WSD highlight the use of unsupervised techniques, pre-trained models, and handling the complexities of multilingual disambiguation. By developing Multilingual WSD we can apply for multiple languages irrespective of language characteristics.

7   Challenges in WSD

The complexity and ambiguity of language pose a variety of difficulties for word sense disambiguation. The following section discusses a few difficulties with WSD.

Ambiguity in Complex Sentences: Complex sentences with multiple ambiguous words pose a formidable challenge for WSD. Some sentences are inherently difficult to disambiguate, even for humans, necessitating the incorporation of additional contextual information. **Contextual Dependency:** The meaning of words depends a lot on the context they're used. This makes Word Sense Disambiguation (WSD) pretty tricky. To get the right meaning, you have to look at the whole sentence or even the entire document. Spotting the tiny shifts in meaning depending on the context is super important for accurate WSD.

Cross-Lingual Disambiguation: When translating WSD information between languages, it is a big challenge because of differences in sense inventories and lack of direct mapping between senses. Cross-lingual disambiguation for translation purposes is difficult but necessary to make sure the language translation is effective and easier to understand. **Sense Granularity and Fine-grained Senses:** One of the significant challenges in WSD is the granularity

Table 8. Summary of WSD Challenges across Languages

Language	Challenges	Unique Characteristics	Open Issues
English	Ambiguity in Polysemous Words, Limited Domain-Specific Data	Morphologically poor, Extensive use of idioms	sense granularity, fine-grained senses, effective contextual incorporation, domain-specific language usage
Chinese	Character level ambiguity, Lack of Clear Word Borders, Contextual Changes and Syntactic Flexibility	Unique character-based writing system(logographic in nature)	Cross-lingual sense alignment, Improving word segmentation, contextual ambiguity
Indian	Lack of Annotated Data, Dialectal Variations, Limited Lexical Coverage	Rich linguistic diversity, Script variations	Development of dialect-specific models, Semantic complexity, Overcoming resource scarcity
Arabic	contextual variations according to diacritical markings, lack of lexical resources	Rich morphological structure	Building standardized evaluation, Cross-lingual and context-aware models
Amharic	Lack of standard dataset and Limited Parallel Corpora	Morphologically rich and the language characterized by a lot of dialect	Development of cross-lingual model and beefing from pre-trained models

of senses, especially when it comes to fine-grained ones. However, differentiating between closely related senses accurately can be difficult since many words have intricate meanings that change depending on the context. On applications like sentiment analysis where slight changes in meaning influence the overall interpretation of text, fine-grained distinctions are necessary. Still, constructing robust models capable of handling fine-grained senses and sense granularity remains problematic in advancing WSD.

Handling Polysemy and Homonymy: Polysemy and homophones refer to words with multiple meanings within one sense or different senses but the same forms, respectively, and they often pose a problem in WSD. Models need to capture intricate contexts and semantic relations to differentiate between different cases. Traditional methods often struggle to achieve accurate disambiguation when polysemy and homonymy occur (Table 8).

8   Conclusion

In this Paper, we have addressed state-of-the-art Word Sense Disambiguation methods across diverse languages, such as English, Chinese, Indian languages, Arabic, and Amharic. Our exploration included a comprehensive discussion and comparison of word sense disambiguation approaches and language-specific challenges, and we highlighted open issues in each context.

We observed the efficacy of pre-trained models and deep learning across languages, emphasizing their potential to overcome challenges. Deep learning has improved WSD through its ability to capture semantic ambiguity and context complexity. However, some deep learning works have shown encouraging results with the use of attention mechanisms; persistent issues include the inability to capture full features of context from sentences, limited data for certain languages, and inherent word ambiguity. Nevertheless, there are difficulties for such mechanisms in fully understanding the characteristic features of a particular sentence. So by developing such context-aware methods, it is time to integrate or apply these word sense disambiguation techniques for different NLP tasks. Our survey contributes valuable insights to advance WSD methods and tackle specific linguistic challenges.

References

1. Abderrahim, M.A., Abderrahim, M.E.A.: Arabic word sense disambiguation for information retrieval. Trans. Asian Low-Res. Lang. Inform. Process. **21**(4), 1–19 (2022)
2. Agirre, E., Edmonds, P.: Word sense disambiguation: Algorithms and applications, vol. 33. Springer Science & Business Media, Oxford, UK (2007)
3. Agirre, E., López de Lacalle, O., Soroa, A.: Random walks for knowledge-based word sense disambiguation. Comput. Linguist. **40**(1), 57–84 (2014)
4. Alian, M., Awajan, A., Al-Kouz, A.: Arabic word sense disambiguation - survey. In: 2017 International Conference on New Trends in Computing Sciences (ICTCS), pp. 236–240 (2017). https://doi.org/10.1109/ICTCS.2017.23
5. Aliwy, A., Taher, H.: Word sense disambiguation: survey study. J. Comput. Sci. **15**, 1004–1011 (2019). https://doi.org/10.3844/jcssp.2019.1004.1011
6. AlMousa, M., Benlamri, R., Khoury, R.: A novel word sense disambiguation approach using wordnet knowledge graph. Comput. Speech Lang. **74**, 101337 (2022)
7. Antoun, W., Baly, F., Hajj, H.: Arabert: Transformer-based model for arabic language understanding. arXiv preprint arXiv:2003.00104 (2020)
8. Assemu, S.: Unsupervised machine learning approach for word sense disambiguation to amharic words. Unpublished Master's Thesis, Department of Information Science, Addis Ababa University, Addis Ababa, Ethiopia (2011)
9. Atwell, E., Al-Sulaiti, L., Al-Osaimi, S., Abu Shawar, B.: A review of arabic corpus analysis tools. In: Proceedings of TALN04: XI conference sur le traitement automatique des langues naturelles, vol. 2, pp. 229–234 (2004)
10. Banerjee, S., Pedersen, T., et al.: Extended gloss overlaps as a measure of semantic relatedness. In: Ijcai, vol. 3, pp. 805–810 (2003)

11. Barba, E., Procopio, L., Navigli, R., et al.: Consec: word sense disambiguation as continuous sense comprehension. In: Proceedings of the 2021 Conference on Empirical Methods in Natural Language Processing, pp. 1492–1503 (2021)
12. Basile, P., Caputo, A., Semeraro, G.: An enhanced lesk word sense disambiguation algorithm through a distributional semantic model. In: Proceedings of COLING 2014, the 25th International Conference on Computational Linguistics: Technical Papers, pp. 1591–1600. Dublin City University and Association for Computational Linguistics, Dublin, Ireland (August 2014)
13. Bevilacqua, M., Navigli, R., et al.: Breaking through the 80% glass ceiling: raising the state of the art in word sense disambiguation by incorporating knowledge graph information. In: Proceedings of the 58th Annual Meeting of the Association for Computational Linguistics, pp. 2854–2864. Association for Computational Linguistics (2020)
14. Borah, P.P., Talukdar, G., Baruah, A.: Approaches for word sense disambiguation-a survey. Inter. J. Recent Technol. Eng. **3**(1), 35–38 (2014)
15. Bounhas, I., Elayeb, B., Evrard, F., Slimani, Y.: Arabonto: experimenting a new distributional approach for building Arabic ontological resources. Int. J. Metadata Semant. Ontol. **6**(2), 81–95 (2011)
16. Bounhas, I., Elayeb, B., Evrard, F., Slimani, Y.: Information reliability evaluation: from Arabic storytelling to computer sciences. J. Comput. Cultural Heritage (JOCCH) **8**(3), 1–33 (2015)
17. Breit, A., Revenko, A., Rezaee, K., Pilehvar, M.T., Camacho-Collados, J.: Wictsv: an evaluation benchmark for target sense verification of words in context. arXiv preprint arXiv:2004.15016 (2020)
18. Brockmann, C., Lapata, M.: Evaluating and combining approaches to selectional preference acquisition. In: 10th Conference of the European Chapter of the Association for Computational Linguistics (2003)
19. Brown, P.F., Della Pietra, S.A., Della Pietra, V.J., Mercer, R.L.: Word-sense disambiguation using statistical methods. In: 29th Annual meeting of the Association for Computational Linguistics, pp. 264–270 (1991)
20. Calvo, H., Rocha-Ramirez, A.P., Moreno-Armendáriz, M.A., Duchanoy, C.A.: Toward universal word sense disambiguation using deep neural networks. IEEE Access **7**, 60264–60275 (2019)
21. Chalabi, A.: Sakhr Arabic-English computer-aided translation system. In: Farwell, D., Gerber, L., Hovy, E. (eds.) AMTA 1998. LNCS (LNAI), vol. 1529, pp. 518–521. Springer, Heidelberg (1998). https://doi.org/10.1007/3-540-49478-2_50
22. Chaplot, D.S., Salakhutdinov, R.: Knowledge-based word sense disambiguation using topic models. In: Proceedings of the AAAI Conference on Artificial Intelligence, vol. 32 (2018)
23. Chasin, R., Rumshisky, A., Uzuner, O., Szolovits, P.: Word sense disambiguation in the clinical domain: a comparison of knowledge-rich and knowledge-poor unsupervised methods. J. Am. Med. Inform. Assoc. **21**(5), 842–849 (2014)
24. Chowdhary, K.: Natural language processing for word sense disambiguation and information extraction. arXiv preprint arXiv:2004.02256 (2020)
25. Collins, M.: Parameter estimation for statistical parsing models: theory and practice of distribution-free methods. In: Bunt, H., Carroll, J., Satta, G. (eds.) New developments in parsing technology. Springer, Cambridge (2004). https://doi.org/10.1007/1-4020-2295-6_2
26. Cui, Y., Che, W., Liu, T., Qin, B., Yang, Z.: Pre-training with whole word masking for Chinese bert. IEEE/ACM Trans. Audio, Speech and Lang. Proc. **29**, 3504–3514 (2021). https://doi.org/10.1109/TASLP.2021.3124365

27. Debili, F., Achour, H., Souissi, E.: La langue arabe et l'ordinateur: de l'étiquetage grammatical à la voyellation automatique. Correspondances **71**, 10–28 (2002)
28. Dereje, S.M., Tesfa, T.Y., Yitbarek, W.T., et al.: Sentence level amharic word sense disambiguation. Am. J. Educ. Technol. **1**(2), 83–87 (2022)
29. Devlin, J., Chang, M.W., Lee, K., Toutanova, K.: Bert: Pre-training of deep bidirectional transformers for language understanding. arXiv preprint arXiv:1810.04805 (2018)
30. Edmonds, P., Kilgarriff, A.: Introduction to the special issue on evaluating word sense disambiguation systems. Nat. Lang. Eng. **8**(4), 279–291 (2002). https://doi. org/10.1017/S1351324902002966
31. Elayeb, B.: Arabic word sense disambiguation: a review. Artif. Intell. Rev. **52**(4), 2475–2532 (2019)
32. ElJundi, O., Antoun, W., El Droubi, N., Hajj, H., El-Hajj, W., Shaban, K.: hulmona: the universal language model in Arabic. In: Proceedings of the Fourth Arabic Natural Language Processing Workshop, pp. 68–77 (2019)
33. Elmougy, S., Taher, H., Noaman, H.: Naïve bayes classifier for arabic word sense disambiguation. In: Proceeding of the 6th International Conference on Informatics and Systems, pp. 16–21. Citeseer (2008)
34. Escudero Bakx, G., Màrquez Villodre, L., Rigau Claramunt, G.: On the portability and tuning of supervised word sense disambiguation systems (2000). https:// upcommons.upc.edu/
35. Fellbaum, C., Miller, G.A.: Morphosemantic links in wordnet. Traitement automatique de langue **44**(2), 69–80 (2003)
36. Gale, W.A., Church, K., Yarowsky, D.: Estimating upper and lower bounds on the performance of word-sense disambiguation programs. In: 30th Annual Meeting of the Association for Computational Linguistics, pp. 249–256 (1992)
37. Gawande, U., Kale, S., Thaokar, C.: A novel approach of word sense disambiguation for marathi language using machine learning. In: Recent Advances in Material, Manufacturing, and Machine Learning, pp. 643–652. CRC Press, Abingdon, Oxon, OX14 4RN (2023)
38. Goldberg, Y.: Neural network methods for natural language processing. Springer Nature, Switzerland (2022). https://doi.org/10.1007/978-3-031-02165-7
39. Gomes, J., Jr., de Mello, R.C., Ströele, V., de Souza, J.F.: A study of approaches to answering complex questions over knowledge bases. Knowl. Inf. Syst. **64**(11), 2849–2881 (2022)
40. Gopal, S., Haroon, R.P.: Malayalam word sense disambiguation using naïve bayes classifier. In: 2016 International Conference on Advances in Human Machine Interaction (HMI), pp. 1–4 (2016). https://doi.org/10.1109/HMI.2016.7449181
41. Habash, N., Rambow, O.: Arabic diacritization through full morphological tagging. In: Human Language Technologies 2007: The Conference of the North American Chapter of the Association for Computational Linguistics; Companion Volume, Short Papers, pp. 53–56 (2007)
42. Habash, N., Roth, R., Rambow, O., Eskander, R., Tomeh, N.: Morphological analysis and disambiguation for dialectal arabic. In: Proceedings of the 2013 Conference of the North American Chapter of the Association for Computational Linguistics: Human Language Technologies, pp. 426–432 (2013)
43. Hadiwinoto, C., Ng, H.T., Gan, W.C.: Improved word sense disambiguation using pre-trained contextualized word representations. arXiv preprint arXiv:1910.00194 (2019)
44. Haffari, G.R., Sarkar, A.: Analysis of semi-supervised learning with the yarowsky algorithm. arXiv preprint arXiv:1206.5240 (2012)

45. Hagerie, W.: Ensemble classifiers applied to amharic word sense disambiguation. Addis Ababa University (2013)

46. Hou, B., Qi, F., Zang, Y., Zhang, X., Liu, Z., Sun, M.: Try to substitute: an unsupervised Chinese word sense disambiguation method based on hownet. In: Proceedings of the 28th International Conference on Computational Linguistics. pp. 1752–1757. International Committee on Computational Linguistics, Barcelona, Spain (Online) (Dec 2020). https://doi.org/10.18653/v1/2020.coling-main.155, https://aclanthology.org/2020.coling-main.155

47. Huang, L., Sun, C., Qiu, X., Huang, X.: Glossbert: Bert for word sense disambiguation with gloss knowledge. arXiv preprint arXiv:1908.07245 (2019)

48. Huang, L., Sun, C., Qiu, X., Huang, X.: Glossbert: Bert for word sense disambiguation with gloss knowledge. In: Proceedings of the 2019 Conference on Empirical Methods in Natural Language Processing and the 9th International Joint Conference on Natural Language Processing (EMNLP-IJCNLP), pp. 3509–3514. Association for Computational Linguistics, Hong Kong, China (Nov 2019). https://doi.org/10.18653/v1/D19-1355, https://aclanthology.org/D19-1355

49. Hung, C., Chen, S.J.: Word sense disambiguation based sentiment lexicons for sentiment classification. Knowl.-Based Syst. **110**, 224–232 (2016)

50. Joachims, T.: Text categorization with support vector machines: learning with many relevant features. In: Nédellec, C., Rouveirol, C. (eds.) ECML 1998. LNCS, vol. 1398, pp. 137–142. Springer, Heidelberg (1998). https://doi.org/10.1007/BFb0026683

51. Jose, R., Chooralil, V.S.: Prediction of election result by enhanced sentiment analysis on twitter data using word sense disambiguation. In: 2015 International Conference on Control Communication & Computing India (ICCC), pp. 638–641 (2015). https://doi.org/10.1109/ICCC.2015.7432974

52. Kang, H., Blevins, T., Zettlemoyer, L.: Translate to disambiguate: Zero-shot multilingual word sense disambiguation with pretrained language models. arXiv preprint arXiv:2304.13803 (2023)

53. Kassie, T.: Word sense disambiguation for amharic text retrieval: A case study for legal documents. Addis Ababa, Ethiopia. Masters Thesis Addis Ababa University, Ethiopia (2009)

54. Kharate, N.G., Patil, V.H.: Word sense disambiguation for marathi language using wordnet and the lesk approach. In: Patil, V.H., Dey, N., N. Mahalle, P., Shafi Pathan, M., Kimbahune, V.V. (eds.) Proceeding of First Doctoral Symposium on Natural Computing Research. LNNS, vol. 169, pp. 45–54. Springer, Singapore (2021). https://doi.org/10.1007/978-981-33-4073-2_5

55. Kilgarriff, A., Palmer, M.: Introduction to the special issue on senseval. Comput. Humanit. **34**, 1–13 (2000)

56. Kolte, S.G., Bhirud, S.G.: Word sense disambiguation using wordnet domains. In: 2008 First International Conference on Emerging Trends in Engineering and Technology, pp. 1187–1191 (2008). https://doi.org/10.1109/ICETET.2008.231

57. Lan, Z., Chen, M., Goodman, S., Gimpel, K., Sharma, P., Soricut, R.: Albert: a lite bert for self-supervised learning of language representations. arXiv preprint arXiv:1909.11942 (2019)

58. Lee, Y.K., Ng, H.T.: An empirical evaluation of knowledge sources and learning algorithms for word sense disambiguation. In: Proceedings of the 2002 Conference on Empirical Methods in Natural Language Processing (EMNLP 2002), pp. 41–48 (2002)

59. Lesk, M.: Automatic sense disambiguation using machine readable dictionaries: how to tell a pine cone from an ice cream cone. In: Proceedings of the 5th Annual International Conference on Systems Documentation, pp. 24–26 (1986)
60. Liu, Y., et al.: Roberta: a robustly optimized bert pretraining approach. arXiv preprint arXiv:1907.11692 (2019)
61. Loureiro, D.A.B.: Learning word sense representations from neural language models (2023). https://repositorio-aberto.up.pt/
62. Lu, W., et al.: Graph-based chinese word sense disambiguation with multi-knowledge integration. Comput. Mater. Continua **61**(1) (2019)
63. Luan, Y., Hauer, B., Mou, L., Kondrak, G.: Improving word sense disambiguation with translations. In: Proceedings of the 2020 Conference on Empirical Methods in Natural Language Processing (EMNLP), pp. 4055–4065. Association for Computational Linguistics, Online (Nov 2020). https://doi.org/10.18653/v1/2020.emnlp-main.332, https://aclanthology.org/2020.emnlp-main.332
64. Luo, F., Liu, T., He, Z., Xia, Q., Sui, Z., Chang, B.: Leveraging gloss knowledge in neural word sense disambiguation by hierarchical co-attention. In: Proceedings of the 2018 Conference on Empirical Methods in Natural Language Processing, pp. 1402–1411 (2018)
65. Luo, F., Liu, T., Xia, Q., Chang, B., Sui, Z.: Incorporating glosses into neural word sense disambiguation. arXiv preprint arXiv:1805.08028 (2018)
66. Mante, R., Kshirsagar, M., Chatur, P.: A review of literature on word sense disambiguation. Int. J. Comput. Sci. Inf. Technol. (IJCSIT) **5**(2), 1475–1477 (2014)
67. McCarthy, D., Koeling, R., Weeds, J., Carroll, J.: Finding predominant word senses in untagged text. In: Proceedings of the 42nd Annual Meeting of the Association for Computational Linguistics (ACL 2004), pp. 279–286. Barcelona, Spain (Jul 2004). https://doi.org/10.3115/1218955.1218991, https://aclanthology.org/P04-1036
68. Mekonen, S.: Word sense disambiguation for amharic text: a machine learning approach. Unpublished Master's Thesis, pp. 1–94 (2010)
69. Miller, G.A.: Wordnet: a lexical database for english. Commun. ACM **38**(11), 39-41 (1995). https://doi.org/10.1145/219717.219748
70. Miller, G.A., Chodorow, M., Landes, S., Leacock, C., Thomas, R.G.: Using a semantic concordance for sense identification. In: Human Language Technology: Proceedings of a Workshop held at Plainsboro, New Jersey, 8-11 March (1994). https://aclanthology.org/H94-1046
71. Moro, A., Raganato, A., Navigli, R.: Entity linking meets word sense disambiguation: a unified approach. Trans. Assoc. Comput. Linguist. **2**, 231–244 (2014)
72. Mossa, N., Meshesha, M.: Amharic sentence-level word sense disambiguation u sing transfer learning. In: Artificial Intelligence and Digitalization for Sustainable Development: 10th EAI International Conference, ICAST 2022, Bahir Dar, Ethiopia, pp. 227–238. Springer (2023). https://doi.org/10.1007/978-3-031-28725-1_14
73. Mulugeta, M.: Word Sense Disambiguation for Amharic Sentences using WordNet Hierarchy. Ph.D. thesis, Bahirdar University (2020)
74. Nanjundan, P., Mathews, E.Z.: An analysis of word sense disambiguation (wsd). In: Proceedings of the International Health Informatics Conference: IHIC 2022, pp. 251–259. Springer (2023). doi: https://doi.org/10.1007/978-981-19-9090-8_22
75. Navigli, R.: Word sense disambiguation: a survey. ACM Comput. Surv. (CSUR) **41**(2), 1–69 (2009)

76. Navigli, R., Ponzetto, S.P.: Babelnet: the automatic construction, evaluation and application of a wide-coverage multilingual semantic network. Artif. Intell. **193**, 217–250 (2012)
77. Nguyen, Q.P., Vo, A.D., Shin, J.C., Ock, C.Y.: Effect of word sense disambiguation on neural machine translation: A case study in korean. IEEE Access **6**, 38512–38523 (2018). https://doi.org/10.1109/ACCESS.2018.2851281
78. Olika, S.: Word Sense Disambiguation for Afaan Oromo: Using Knowledge Base. Ph.D. thesis, St. Mary's University (2018)
79. Onyshkevych, B.A.: An ontological-semantic framework for text analysis. Ph.D. thesis, Carnegie Mellon University (1997)
80. Pal, A.R., Kundu, A., Singh, A., Shekhar, R., Sinha, K.: A hybrid approach to word sense disambiguation combining supervised and unsupervised learning. arXiv preprint arXiv:1611.01083 (2015)
81. Pal, A.R., Saha, D.: Word sense disambiguation: A survey. arXiv preprint arXiv:1508.01346 (2015)
82. Peters, M.E., Neumann, M., Iyyer, M., Gardner, M., Clark, C., Lee, K., Zettlemoyer, L.: Deep contextualized word representations (2018)
83. Poibeau, T.: Machine translation. MIT Press, Cambridge (2017)
84. Priss, U., Old, L.J.: Revisiting the potentialities of a mechanical thesaurus. In: Ferré, S., Rudolph, S. (eds.) ICFCA 2009. LNCS (LNAI), vol. 5548, pp. 284–298. Springer, Heidelberg (2009). https://doi.org/10.1007/978-3-642-01815-2_21
85. Priya, M.S., Renuka, D.K., Kumar, L.A., Rose, S.L.: Multilingual low resource indian language speech recognition and spell correction using indic bert. Sādhanā **47**(4), 227 (2022)
86. Raganato, A., Bovi, C.D., Navigli, R.: Neural sequence learning models for word sense disambiguation. In: Proceedings of the 2017 Conference on Empirical Methods in Natural Language Processing, pp. 1156–1167 (2017)
87. Raganato, A., Camacho-Collados, J., Navigli, R., et al.: Word sense disambiguation: a uinified evaluation framework and empirical comparison. In: Proceedings of the 15th Conference of the European Chapter of the Association for Computational Linguistics: Volume 1, Long Papers, vol. 1, pp. 99–110 (2017)
88. Resnik, P.: A perspective on word sense disambiguation methods and their evaluation. In: Tagging Text with Lexical Semantics: Why, What, and How? (1997)
89. Rothman, D.: Transformers for Natural Language Processing: Build innovative deep neural network architectures for NLP with Python, PyTorch, TensorFlow, BERT, RoBERTa, and more. Packt Publishing Ltd, Birmingham, UK (2021)
90. Saeed, A., Nawab, R.M.A., Stevenson, M.: Investigating the feasibility of deep learning methods for urdu word sense disambiguation. Trans. Asian Low-Resource Lang. Inform. Process. **21**(2) (2021). https://doi.org/10.1145/3477578
91. Saeidi, M., Mahdaviani, K., Milios, E., Zeh, N.: Context-enhanced concept disambiguation in wikification. Intell. Syst. Appli.. 200246 (2023)
92. Sarmah, J., Sarma, S.K.: Decision tree based supervised word sense disambiguation for assamese. Int. J. Comput. Appl. **141**(1), 42–48 (2016)
93. Seneff, S.: TINA: A natural language system for spoken language applications. Comput. Linguist. **18**(1), 61–86 (1992). https://aclanthology.org/J92-1004
94. Singh, H., Bhattacharyya, P.: A survey on word sense disambiguation. ACM Comput. Surv. (CSUR) (2019)
95. Singh, R.L., Ghosh, K., Nongmeikapam, K., Bandyopadhyay, S.: A decision tree based word sense disambiguation system in Manipuri language. Adv. Comput. **5**(4), 17 (2014)

96. Song, Y., Ong, X.C., Ng, H.T., Lin, Q.: Improved word sense disambiguation with enhanced sense representations. In: Findings of the Association for Computational Linguistics: EMNLP 2021, pp. 4311–4320 (2021)

97. Stevenson, M., Wilks, Y.: The interaction of knowledge sources in word sense disambiguation. Comput. Linguist. **27**(3), 321–349 (2001)

98. Stokoe, C., Oakes, M.P., Tait, J.: Word sense disambiguation in information retrieval revisited. In: Proceedings of the 26th annual international ACM SIGIR Conference on Research and Development in Informaion Retrieval, pp. 159–166 (2003)

99. Sumanth, C., Inkpen, D.: How much does word sense disambiguation help in sentiment analysis of micropost data? In: Proceedings of the 6th Workshop on Computational Approaches to Subjectivity, Sentiment and Social Media Analysis, pp. 115–121 (2015)

100. Sun, X.R., Lv, S.H., Wang, X.D., Wang, D.: Chinese word sense disambiguation using a lstm. In: ITM Web of Conferences, vol. 12, p. 01027. EDP Sciences (2017)

101. Tesema, W., Tesfaye, D., Kibebew, T.: Towards the sense disambiguation of afan oromo words using hybrid approach (unsupervised machine learning and rule based). Ethiopian J. Educ. Sci. **12**(1), 61–77 (2016)

102. Vial, L., Lecouteux, B., Schwab, D.: Sense vocabulary compression through the semantic knowledge of wordnet for neural word sense disambiguation. arXiv preprint arXiv:1905.05677 (2019)

103. Vickrey, D., Biewald, L., Teyssier, M., Koller, D.: Word-sense disambiguation for machine translation. In: Proceedings of Human Language Technology Conference and Conference on Empirical Methods in Natural Language Processing, pp. 771–778 (2005)

104. Wang, Y., Wang, M., Fujita, H.: Word sense disambiguation: a comprehensive knowledge exploitation framework. Knowl.-Based Syst. **190**, 105030 (2020)

105. Wang, Y., Zheng, K., Xu, H., Mei, Q.: Clinical word sense disambiguation with interactive search and classification. In: AMIA Annual Symposium Proceedings, vol. 2016, p. 2062. American Medical Informatics Association (2016)

106. Wassie, G., Ramesh, B., Teferra, S., Meshesha, M.: A word sense disambiguation model for amharic words using semi-supervised learning paradigm. Sci. Technol. Arts Res. J. **3**(3), 147–155 (2014)

107. Weaver, W.: Information theory, p. 232. eM Publications (1949)

108. Wilks, Y.: A preferential, pattern-seeking, semantics for natural language inference. Artif. Intell. **6**(1), 53–74 (1975)

109. Wilks, Y., Fass, D., Guo, C.M., McDonald, J.E., Plate, T., Slator, B.M.: Providing machine tractable dictionary tools. Mach. Transl. **5**, 99–154 (1990)

110. Wu, Y., Jiang, M., Xu, J., Zhi, D., Xu, H.: Clinical named entity recognition using deep learning models. In: AMIA Annual Symposium Proceedings, vol. 2017, p. 1812. American Medical Informatics Association (2017)

111. Yarowsky, D.: Unsupervised word sense disambiguation rivaling supervised methods. In: 33rd Annual Meeting of the Association for Computational Linguistics, pp. 189–196. Association for Computational Linguistics, Cambridge, Massachusetts, USA (Jun 1995). https://doi.org/10.3115/981658.981684, https://aclanthology.org/P95-1026

112. Yarowsky, D., Florian, R.: Evaluating sense disambiguation across diverse parameter spaces. Nat. Lang. Eng. **8**(4), 293–310 (2002)

113. Z., Y., H., H.: Graph based word sense disambiguation method using distance between words. J. Softw. **23**(4), 776–785 (2012)

114. Zhang, X., Hauer, B., Kondrak, G.: Improving hownet-based chinese word sense disambiguation with translations. In: Findings of the Association for Computational Linguistics: EMNLP 2022, pp. 4530–4536. Association for Computational Linguistics, Abu Dhabi, United Arab Emirates (Dec 2022), https://aclanthology.org/2022.findings-emnlp.331

115. Zhong, Z., Ng, H.T.: Word sense disambiguation improves information retrieval. In: Proceedings of the 50th Annual Meeting of the Association for Computational Linguistics (Volume 1: Long Papers), pp. 273–282 (2012)

116. Zitouni, I.: Natural language processing of semitic languages. Springer, New York (2014). https://doi.org/10.1007/978-3-642-45358-8

Ge'ez Syntax Error Detection Using Deep Learning Approaches

Habtamu Shiferaw Asmare[1,2]([✉])[iD] and Abdulkerim Mohammed Yibre[1][iD]

[1] Department of Information Technology, Faculty of Computing, Bahir Dar Institute of Technology, Bahir Dar University, Bahir Dar, Ethiopia
abdulkerim.mohamed@bdu.edu.et
[2] Department of Information Technology, College of Computing and Informatics, Mekdela Amba University, Tulu Awulia, Ethiopia
habtamu_shiferaw@mkau.edu.et

Abstract. The Ge'ez language, an ancient Ethiopian Semitic language, is still used in liturgical contexts and taught at university and college levels but lacks tools for part of speech tagging, morphological analyzers, and syntax error detection in written texts. This hinders the identification of syntax errors and poses a significant challenge for learners, and researchers. This study addresses this problem by developing a part of speech tagging and syntax error detection models for Ge'ez using deep learning approaches. To develop the model, a dataset of 4,981 sentences that have 30326 words and 11,747 unique words was collected for part of speech tagging. Additionally, a dataset of 1,170 sentences was collected for syntax error detection. LSTM and BiLSTM algorithms were used to develop the models. The LSTM model achieved an accuracy of 94% and 92.31% in the Gz_POS and Gz_SED tasks, respectively, and the BiLSTM model achieved an accuracy of 95.01% and 94.02% in the Gz_POS and Gz_SED. The results demonstrate the effectiveness of these deep learning algorithms for syntax error detection in the Ge'ez language. The developed model provides a feasible solution to the challenges of digitizing Ge'ez books and provides help for second-language learners. The findings contribute to the improvement of language education, research, and development in under-resourced languages. Future researchers can use the developed model and methodology as a framework for further advancements in Ge'ez language processing.

Keywords: Ge'ez POS Tagging · Ge'ez Syntax Error Detection · Deep Learning · LSTM · BiLSTM

1 Introduction

In recent years, there has been growing interest in learning Ge'ez outside of Ethiopia and Eritrea within the universities and online learning sites. To improve the quality of written text in word processing software, emails, blog posts, and

T. G. Debelee et al. (Eds.): PanAfriConAI 2023, CCIS 2068, pp. 203–220, 2024.
https://doi.org/10.1007/978-3-031-57624-9_11

chats, grammar syntax error detection models need to be developed for writers. These models are developed to follow a set of rules defining what correct sentences should look like, enabling writers to assess sentence correctness [10]. Various approaches, such as pattern-matching, syntax-based, rule-based, statistical, deep learning, and hybrid methods, are employed in developing the syntax error detection model [1, 22].

Considering the significance of the Ge'ez language and the need for NLP (Natural Language Processing) applications specific to Ge'ez, syntax error detection, and part of speech tagging models are developed for the Ge'ez language. By developing such models, the understanding and processing of the language will be further advanced, benefiting the preservation and analysis of historical, philosophical, ethical, religious, and other indigenous literature written in Ge'ez.

Ge'ez holds significance as one of the major literary languages in the ancient world, with a history spanning two millennia in the Horn of Africa and Arabia. Despite its historical and cultural value, Ge'ez remains under-resourced in terms of language technology tools and is listed under minority languages that do not get attention in Ethiopian natural language processing [24]. Therefore, the development of a deep learning-based syntax error detection with part of speech tagging for Ge'ez has a significant impact on the Ge'ez language-speaking community and contributes to the broader field of natural language processing. Despite the existence of numerous books, literature, and archives in Ge'ez, the process of digitizing them often results in syntax errors.

Additionally, Ge'ez writers make syntax errors in document preparation while they write Ge'ez texts in word processing software [1] page 16. This hinders the identification of syntax errors and poses a significant challenge for learners, researchers, and developers working with the language.

Currently, no tool developed to identify the POS(part of speech) tags, Morphologies for Ge'ez words, and syntax errors for Ge'ez sentences. This exacerbates the resource limitations and hinders the development of higher-level NLP applications such as syntax error correction, machine translation, semantic analysis, and text summarization. While syntax error detection models have been developed for various languages, including English [25], Tigrigna [11], Afan Oromo [22], and Amharic [5,20]. The unique characteristics of the Ge'ez language, such as its free word order for subject, object, and verb combinations (svo, sov, osv, ovs, vso, or vos), present challenges for adapting existing syntax error detection models from another language.

Therefore, a dedicated syntax error detection model specifically designed for Ge'ez is required to support learners, researchers, and developers in producing error-free texts and preparing reliable datasets for Ge'ez language processing tasks.

Previous research efforts in Ge'ez language processing, including Ge'ez machine translation [1,2], next word prediction [17], factoid question answering [4], and information extraction [23], have identified the need for syntax error detection to enhance the quality and reliability of their datasets. However, no existing studies have focused on syntax error detection in Ge'ez. The motivation stems from the desire to bridge the technological gap in Ge'ez language resources, improve language learning and writing skills, and contribute to the preservation and advancement of the Ge'ez language within the context of natural language processing. Our contributions are summarized as follows.

- We collected 4981 sentences having 30326 words with 11,747 unique words for part of speech tagging and 1170 sentences for syntax error detection model development.
- We annotated the part of speech dataset with 84 unique part of speech tags.
- We developed a part of speech tagger that achieves a state-of-the-art result in the Ge'ez part of speech tagging using deep learning compared to [13].
- We developed a Ge'ez syntax error detection model using LSTM and BiLSTM that was not previously tried. So, this paper can be a role model or a framework for upcoming researchers in the field.

The paper contributes towards tackling the challenge of digitizing books, literature, and archives written in the Ge'ez language by developing a deep learning-based syntax error detection model. It also contributes to bridging the gap in Ge'ez language processing tools and provides valuable insights for learners, researchers, and developers working with the language.

The rest of the paper includes the following sections. Section two focuses on the discussion of the language's part of speech and morphology, as well as exploring related literature. Section three provides insight into the methodology, which includes details on the model architecture, dataset collection, preprocessing, model development, and deep learning-based syntax error detection. The fourth section showcases the experimental results of the model, highlighting the hyper-parameter setups, evaluation metrics, model results, overall model accuracy comparison, and sample predictions. Lastly, the conclusion and future works are presented in the last section.

2 Ge'ez Language

In this section, the Ge'ez language is presented. It contains an introduction to the Ge'ez writing system, Ge'ez part of speech, Ge'ez morphology, and types Ge'ez syntax errors. Additionally, this section provides a detailed discussion of related research on detecting syntax errors in local and foreign languages.

2.1 Ge'ez Writing System

Ge'ez script or the Ethiopic script is the writing system used for the Ge'ez language. It is an abugida script that was originally developed to write the ancient Semitic language of Ge'ez, which is still used as the liturgical language of the Ethiopian Orthodox Tewahedo Church(EOTC). The Ge'ez script is written from left to right [1] and it conveys grammatical information through affixes, such as prefixes, infixes, and suffixes, attached to the roots or stems. In Western contexts, Ge'ez is sometimes called "Old Ethiopic/Classical Ethiopic" or simply "Ethiopic." The terms Ethiopic and Ge'ez are often used interchangeably [7,15,19].

2.2 Ge'ez Part of Speech

Part of speech refers to the grammatical category into which words are classified based on their functions and roles within sentences. In Ge'ez, there are eight recognized part of speech: noun, pronoun, verb, adjective, adverb, preposition, conjunction, and interjection. A general description of the Ge'ez part of speech can be found in our previously published paper [3].

2.3 Ge'ez Morphology

Morphology is the branch of linguistics that studies the structure of words and the varying patterns of how words are formed. Morphology deals with the study of inflectional and derivational forms of words. Inflectional forms of words indicate their grammatical categories, such as tense, number, case, and gender, while derivational forms change the meaning of the word by adding prefixes, suffixes, or infixes [14].

There are two productive ways to create words from morphemes: inflection and derivation. Hence, there are two types of morphology: inflectional and derivational. Inflectional morphology involves combining a word stem with a grammatical morpheme, resulting in a word of the same part of speech as the original, and often serving a syntactic function. For example, Ge'ez uses the morpheme /-kmu/ to mark the 2nd person plural masculine subject.

Inflectional morphemes affect the grammatical function of the stem but not its part of speech. Inflection produces different surface forms of the same word. A given transitive verb in Ge'ez can produce more than 1,388 inflectional surface forms [6].

In contrast, derivation morphology involves combining a word root with a grammatical morpheme that often results in a word of a different part of speech, with a meaning that may be difficult to predict precisely. Ge'ez has derivational affixes that can change the part of speech of a word. For example, the verb (*fenewe*/he sent) can be transformed into the adjective (*fenawi*/sender) by adding the derivational suffix /-i/ [6]. We considered four types of morphological inflection of words in this work due to time and resource constraints. These are inflections due to number (singular or plural), gender (feminine or masculine), tense (past, future, present, imperative, or gerundive), and person (1st, 2nd, or 3rd).

Verbs can be inflected to number, gender, tense, and person. Adverbs can be inflected to number, gender, tense, or time. Adjectives and Nouns can be inflected to number and gender. For example, the verb (*qetele*/he killed) is past tense, masculine, past tense, and 3rd person. This word can be inflected to (*yiqetl*/he kills) is future, (yiqtl/to kill) is present, (*yiqtl*/kill) is imperative, and (*qetil* or *qetilot*/killing) is gerundive where all these are masculine and 3rd person. Additionally, (*qetelet*/she killed) is feminine 3rd person singular, and (*qetelu*/they killed) is masculine 3rd person plural.

2.4 Types of Ge'ez Syntax Errors

There are about 17 (seventeen) syntax errors in Geez, but the most commonly occurring syntax errors include numerical disagreement, gender disagreement, and time/tense disagreement [18].

Numerical Disagreement occurs when there is an error in the agreement between a word and a numeric quantifier such as a number or a counting word. If a writer uses a singular verb with a plural noun or a plural verb with a singular noun, this would result in a numerical disagreement. Example, (*menu wuetu be'esi zeyfekdu haywe*/Who wants to be saved). In this sentence, the word (*zeyfekdu*) is for plural but the word (*menu*) indicates a singular subject. This turns the sentence into a numerical disagreement.

Gender disagreement occurs when there is a disagreement between the gender of the subject and the gender of a pronoun used to refer to that subject. For example, (*itkuni keme feres webekl ele albomu lib*/Don't be like a horse and a mule that have no heart). Here the word (*itkuni*) indicates singular feminine while the original text was to say (*itkunu*) which is plural masculine resulting in gender disagreement.

Time or tense disagreement, on the other hand, occurs when there is an agreement issue between the tenses used in a sentence. If a writer uses a past tense when referring to a present or future event, this would result in time disagreement. For example, (*hye wodku kulomu geberte ametsa, yisededu weiyiklu qewime*/And all that do iniquity fall from it; They have fallen, and cannot stand). From this sentence, the word (*wodku*) indicates past tense while (*yisededu*) indicates future tense resulting in time disagreement between the two verbs.

2.5 Related Works

Several studies have been conducted in different foreign languages in the field of syntax error detection. For example, in English [25], in the Indonesian language [21], and in Chinees [12,16]. Numerous studies on detecting syntax errors have also been conducted in different Ethiopian languages.

Yared [5] developed a deep learning model to detect grammar errors in Amharic. They used the HornMorpho tool for Amharic word analysis and trained the model on a dataset consisting of 3882 sentences. The model, based on LSTM and BiLSTM, achieved accuracies of 88.27% and 88.89% respectively. Although Amharic is related to Ge'ez, the findings cannot be directly applied to Ge'ez syntax error detection.

Another work by Molaligne [20] was developed using deep learning to detect and correct grammar syntax errors in Amharic. The model used POS tagging, named entity recognition, and morphological analysis to analyze Amharic words. They used 8.25K and 6.69K Amharic sentences for detection and correction respectively. LSTM and BiLSTM were employed for error detection, achieving accuracies of 65% and 80%, respectively. An Encoder-Decoder model with and without Attention was used for error correction, achieving a BLEU score of 58.59%. To improve the model, the researchers suggested a larger annotated corpus, the inclusion of multi-word expressions, and word sense disambiguation. Further research can be done to enhance the study's findings.

Gidey [11] also proposed the application of dependency-based grammar-checking for the Tigrinya language. The model comprises a text preprocessing module, a language dependency model, a dependency extraction module, and a grammar-checking module. Evaluation using a test dataset of 122 sentences showed a precision of 92.46%, accuracy of 92.09%, and recall of 61.21%. However, the low recall score indicates difficulty in accurately identifying grammatical errors, particularly false negatives. Additionally, a dependency-based grammar checker depends on predefined linguistic rules, which do not capture different language characters.

Additionally, Tesfaye [22] presents the development and testing of a rule-based Afan Oromo grammar checker, showing promising performance. The checker primarily struggles with compound, complex, and compound-complex sentences due to the focus on simple sentence rules. Additional rules addressing these sentence types can enhance the checker's performance. Some grammatically incorrect sentences are not flagged, potentially due to incomplete rules and incorrectly tagged words from the part of speech tagger component. However, the approach has a limited ability to handle complex language phenomena and adapt to language variations. Additionally, as they described, the rules were not complete and there were incorrectly tagged words in their dataset.

A more recent work by Feleke [8] uses a morphology-based approach to create a Ge'ez spelling error checker for Ge'ez language. The system employed various techniques such as dictionary lookup, morphological analysis, replacement rules, and Levenshtein Edit Distance (LED) for error detection and correction. Using the open-source tool Hun spell, a dictionary consisting of 6115 words and 955 defined rules was constructed. The system achieved a commendable performance of 90.05% based on metrics like recall, precision, and suggestion adequacy. However, the system does not detect syntax errors and the development of well-structured rules is mandatory to further enhance its performance. Additionally, the system currently is not publicly available.

A work by Kassahun & Fantaye [13] used a deep learning approach to develop POS tagging for Ge'ez using a corpus of 2,552 Bible sentences with 26,607 total words. They used 16 tag sets. The BiGRU model achieved the highest accuracy at 86.70%, while the GRU model had an accuracy of 86.15%. The study found that the deep learning approach outperformed traditional methods for developing a POS tagger for low-resource languages like Ge'ez. However, they do not consider the corresponding morphology of words in Ge'ez. Additionally, the tagsets used to prepare the dataset are only 16. This is more general and difficult for deep learning algorithms to identify grammatical syntax errors in morphologically complex languages like Ge'ez. Therefore, we developed our own POS tagger model for Ge'ez [3] that has 84 tagsets with better performance and a higher dataset.

Summary

The lack of a publicly available structured dataset and appropriate linguistic tools for Ge'ez poses challenges in developing spelling and grammar checkers for the language. The significance of spelling and grammar checkers in corpus preparation was highlighted by [4,23]. Several studies have been conducted on syntax error detection in various foreign languages, such as English, Indonesian, and Chinese. Some studies focused on detecting grammar errors in Amharic [5, 20], using deep learning models trained on datasets of Amharic sentences. These models achieved accuracy ranging from 65% to 88.89% but faced challenges related to lexical ambiguity. Another work proposed dependency-based grammar checking for the Tigrinya language [11], achieving high precision and accuracy but struggling with recall. An Afan Oromo grammar checker showed promising performance but had limitations in handling complex language phenomena and adapting to language variations. Recent work on Ge'ez spelling error checking [8] achieved good results but did not address syntax errors and required the development of structured rules. There is no previous work on Ge'ez syntax error detection. While existing studies offer insights into grammar error detection in specific languages, they do not directly address the specific challenges of Ge'ez. Therefore, there is a research gap that calls for the development of a syntax error detection model made explicitly for Ge'ez, considering its unique grammar rules and structures.

3   Materials and Methods

The research design method that we used for deep learning-based grammar syntax error detection model development is the experimental research design. This is because the development of a deep learning model involves the manipulation of independent variables (such as the number and size of layers, activation functions, learning rates, etc.) to see how they affect the dependent variable (such as the model's accuracy in detecting syntax errors in the text), and to establish a causal relationship between them.

Experimental research design allows for the control of other extraneous variables that might affect the dependent variable, and it allows researchers to generalize findings to the population. Therefore, the experimental research design is an appropriate research design for developing a deep learning-based grammar syntax error detection model.

3.1   Dataset Collection and Preparation

Data is collected from different books available online. The data sources include different Holy Bible books found from online repositories, Ge'ez course handouts, and Psalm prayer books. There is similar data available from the online repositories and an error from one source is also an error in the other. This is due to the usage of similar sources.

A total of 4981 sentences were collected with 30K words having 11,741 unique words for developing a part of speech tagger. Additionally, 1170 correct and incorrect sentences were collected for syntax error detection model development. Among these, 335 have a numerical disagreement, 286 have a gender disagreement, 282 have a time disagreement, and 267 are correct sentences. The prepared annotation guideline is based on the syntax of the language. For this, we used the Ge'ez books [7, 15, 18, 19].

The dataset is annotated in collaboration with domain experts and Deacons who know Ge'ez such as poetry students in Bahir Dar Poly Gibi Gubae Poetry School students and the poetry teacher. After acquiring the dataset, the domain experts reviewed the correctness of the labels and approved the dataset after several discussions and mutual agreement with its correct label.

3.2 Model Architecture

The proposed model, as indicated in Fig. 1, for Ge'ez language syntax error detection consists of several components: preprocessing, word representation, model development, and model testing. The input is an annotated dataset that is preprocessed through tokenization, tag generation, encoding or vectorizing, and sequence padding. Word embedding is then used to convert the text into numeric vectors. The model for grammar syntax error detection uses LSTM and BiLSTM algorithms and is evaluated using performance metrics. The model identifies the error types as number disagreement, gender disagreement, time disagreement, or as a correct sentence.

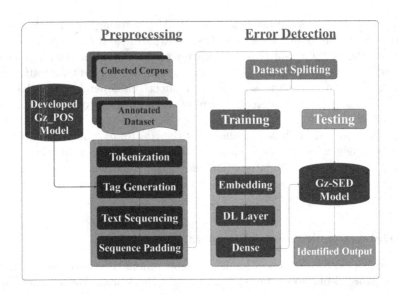

Fig. 1. The detailed architecture of Ge'ez Syntax Error Detection model.

3.3 Text Preprocessing

Text preprocessing refers to the initial phase of text data preparation, where raw text is transformed into a cleaner and more structured format suitable for further analysis or natural language processing tasks. It involves a series of steps intended to prepare the text data valid input for the deep learning algorithm. In the text preprocessing section tokenization, tag generation, adding the tagged values to the dataset, sequencing, and sequence padding are included.

3.4 Model Development

We defined a Long Short-Term Memory (LSTM) and a Bidirectional Long-Short Term Memory (BiLSTM) neural network model for syntax error detection. Since LSTM and BiLSTM are designed to capture long-term dependencies in sequential data, such as sentences in natural language text. Here is a brief description of the used layers and parameters:

Embedding layer: This layer maps each word in the input sequence to a dense vector of fixed size.

LSTM layer: This is the main layer of the LSTM model, which processes the sequence data. The first argument specifies the number of LSTM units (or cells) in the layer.

Dense layer: This is a fully connected layer that produces the final output of the model. The specified argument tells the number of output classes that we used, which is 4 in this case.

4 Experiment and Result

4.1 Dataset Splitting

Two experiments are conducted by different train-validation-test splitting techniques. Experiment 1 takes the 70-15-15 train-validation-test split. A total of 4981 sentences and 1170 sentences are used for POS tagger and syntax error detection model development.

Table 1. Dataset split for Gz_POS.

Experiment	Training	Validation	Testing
Exp1	3598	635	748
Exp2	4033	449	499

From Table 1, for Exp1, there are 3598 samples in the training set, 635 in the validation set, and 748 in the testing set. For Exp2, there are 4033 samples in the training set, 449 in the validation set, and 499 in the testing set. This makes a total of 4981 sentences.

Table 2. Dataset split for Gz_SED.

Experiment	Training	Validation	Testing
Exp1	844	150	176
Exp2	947	106	117

From Table 2, for Exp1, there are 844 samples in the training set, 150 in the validation set, and 176 in the testing set. For Exp2, there are 947 samples in the training set, 106 in the validation set, and 117 in the testing set. This makes a total of 1170 sentences.

Tables 1 and 2 provide information about the split of a dataset for each experiment(Exp1 and Exp2), with different categories and their corresponding values. It summarizes the numbers for the train, validation, and test sets used in the model development process.

4.2   Hyper-parameter Tuning

Parameter tuning: is the process of finding the optimal values of parameters in a machine learning model that help achieve the best performance. It is a critical step in model development and can be time-consuming and challenging.

To reduce the time and resources required for parameter tuning, we used DEAP (Distributed Evolutionary Algorithms in Python) which is a popular open-source framework for implementing evolutionary algorithms in Python [9]. We run the DEAP algorithm with 20 populations and 10 generations to find the best hyper-parameters (Table 3).

Table 3. DEAP hyper-parameters ranges.

No	Hyper-parameter	Ranges
2	Batch Size	8–64
5	Embedding Dimension	10–100
6	Units	32–64
7	Dropout	0.2–0.5
8	Optimizer	Adam, SGD, RMSprop

The DEAP optimization algorithm finds all the possible values from the given ranges of values and selects the best hyper-parameter based on the result histories registered while searching and testing the model. The parameter tuning process takes 10 h for the POS tagger model and 1 h for the syntax error detection model. Using the outputs of the DEAP best hyper-parameters, we got the best fitness of 95.0% and an average fitness of 88.0%. The hyper-parameters, including batch size, dropout rate, embedding dimension, optimizer, and the number of units are taken from the DEAP optimization result. Epoch is based on an early-stopping mechanism (Table 4).

Table 4. DEAP-based and manually set summary of hyper-parameters used for Gz_SED during a model training process

No	Hyper-parameter	LSTM, BiLSTM Values
1	Epoch	20, 18
2	Batch Size	15
3	Sequence Length	10
4	Activation Function	SoftMax
5	Embedding Dimension	27
6	Units	46
7	Dropout	0.42
8	Optimizer	Adam

4.3 Experimental Result

This section presents the part of speech tagger and the syntax error detection models accuracy, loss, confusion matrix, and overall model results.

Gz_POS Tagger Accuracy and Loss: In Fig. 2, the performance metrics of the BiLSTM-based part-of-speech tagger model are visually represented, showing the model's accuracy and loss throughout the training process. Notably, the training and validation accuracy display a consistent upward trend, indicating the model's ability to learn and generalize well from the dataset.

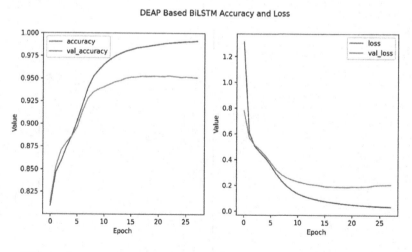

Fig. 2. Accuracy and loss of BiLSTM-based Gz_POS tagger model

A remarkable observation is a persistent increase in training accuracy, persisting even in the final epoch of training. However, the validation accuracy reaches a plateau after the 23rd epoch. Given the early-stopping mechanism's patience parameter set to 5, the model undergoes training until the 28th epoch without any further enhancement in validation accuracy. Consequently, the early-stopping mechanism intervenes, terminating the training process and preserving the model's best performance.

The conclusion of these training efforts results in an impressive accuracy recorded at 95.01% for the BiLSTM model, marking a substantial improvement when compared to its LSTM-based counterpart. Additionally, the model demonstrates a notable reduction in loss, achieving a value of 1.99. This loss reduction signifies the model's improved ability to minimize errors and make more accurate predictions compared to the LSTM model.

Gz_SED Model Accuracy and Loss: Throughout the training process, characterized by multiple epochs, both models demonstrated a consistent decrease in loss values. This decline indicates the models' capacity to iteratively refine their parameters and enhance their understanding of the underlying patterns present in the training data.

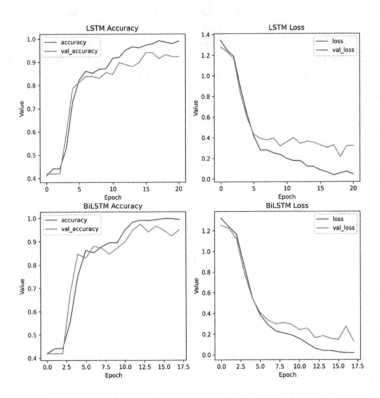

Fig. 3. Accuracy and loss of LSTM and BiLSTM for syntax error detection

In Fig. 3, the LSTM model exhibited a notable validation accuracy of 92.31%, a testament to its proficiency in making accurate predictions on a validation dataset. In comparison, the BiLSTM model surpassed this performance, achieving a higher validation accuracy of 94.02%. The noticeable improvement in accuracy suggests that the bidirectional nature of the BiLSTM architecture contributes significantly to its effectiveness in capturing complex dependencies within sequential data.

Gz_SED Model Confusion Matrix: The analysis presented in Fig. 4 focused on evaluating the performance of both LSTM and BiLSTM models using a dataset containing 117 sentences. These sentences were systematically categorized into various groups, including correct sentences, number disagreements, gender disagreements, and time disagreements. From this, 49 sentences were identified as correct, while 68 sentences showed syntax errors.

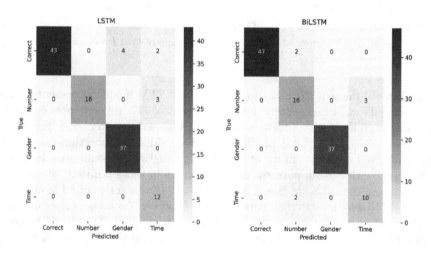

Fig. 4. Confusion matrix of LSTM and BiLSTM for syntax error detection

In the realm of correct sentences in Fig. 4, the LSTM model accurately identified 43 out of 49 instances but misclassified 6 sentences as incorrect. On the other hand, the BiLSTM model demonstrated a slightly higher accuracy by correctly identifying 47 out of 49 correct sentences, with only 2 misclassifications.

When addressing number disagreements, both LSTM and BiLSTM models exhibited similar performance, correctly identifying 16 out of 19 sentences while misclassifying 3 instances. In the case of gender disagreements, both models performed exceptionally well, correctly identifying all 37 sentences with gender-related discrepancies. Regarding time disagreements, the LSTM model achieved a perfect identification rate by correctly recognizing all 12 sentences with time-related issues. The BiLSTM model, while successful in identifying 10 out of 12 sentences with time disagreements, misclassified 2 instances.

Table 5 displays the overall statistics, summarizing their accuracies and misclassifications across different sentence categories. This detailed analysis provides insights into the strengths and weaknesses of both LSTM and BiLSTM models in handling syntactic variations and disagreements within the given dataset.

Table 5. Performance of LSTM and BiLSTM

LSTM	BiLSTM
Total Correct Sentences: 49	Total Correct Sentences: 49
Total Number Disagreement: 37	Total Number Disagreement: 37
Total Gender Disagreement: 19	Total Gender Disagreement: 19
Total Tense/Time Disagreement: 12	Total Tense/Time Disagreement: 12
Total Predictions Made: 117	Total Predictions Made: 117
Total Correct Predictions: 108	Total Correct Predictions: 110
Total Lost Predictions: 9	Total Lost Predictions: 7
Percentage of Correct Predictions: 92.31%	Percentage of Correct Predictions: 94.02%
Percentage of Lost Predictions: 7.69%	Percentage of Lost Predictions: 5.98%

In Table 5, the performance of LSTM and BiLSTM models is presented. The LSTM-based model successfully detected 108 out of 117 test sentences correctly, resulting in an accuracy of 92.31%. The model missed 9 sentences, representing a loss of 7.69% of the total predictions. The results indicate that the LSTM model achieved a high accuracy of 92.31% in predicting the sentences. However, it struggled with certain patterns or complexities in those 9 missed sentences.

Regarding the BiLSTM-based model, out of 117 test sets, BiLSTM correctly identified 110 sentences to their correct types. However, 7 sentences were misidentified, resulting in an accuracy of 94.02% and a loss of 5.98%. The BiLSTM model demonstrated effectiveness in identifying error types in the majority of the test sets.

Tables 6 and 7 present the performance metrics for the LSTM and BiLSTM models in two different tasks and two different experiments for Ge'ez part of speech tagging and Ge'ez syntax error detection. In experiment 1, the models are trained with a 70-15-15 train-validation-test split, and in experiment 2, the models are trained with an 80-10-10 train-validation-test split.

Table 6. Overall model accuracy of Gz_POS tagger in two experiments

Model	Experiment 1	Experiment 2
LSTM	93.76%	94%
BiLSTM	94.5%	95.01%

Table 7. Overall model accuracy of Gz_SED in two experiments

Model	Experiment 1	Experiment 2
LSTM	73.86%	92.31%
BiLSTM	72.72%	94.02%

From the Tables 6 and 7, for the Gz_POS Task LSTM model achieved an accuracy of 93.76% and the corresponding BiLSTM achieved 94.5% in experiment 1. The accuracy of LSTM achieved 94% and BiLSTM achieved 95.01% in experiment 2. For the Gz_SED Task LSTM model achieved an accuracy of 73.86% and the corresponding BiLSTM achieved 72.72% in experiment 1. The accuracy of LSTM is 92.31% and BiLSTM achieved 94.02% in experiment 2.

5   Discussion

The models performed well in identifying sentences. However, they exhibited some mis-classifications. The misclassifications may stem from the ambiguity present in sentences, where multiple types of disagreements can coexist within a sentence. The models struggle to accurately identify the specific disagreement due to the bias of the sentence labeling towards a single disagreement. To improve the models' performance, it is recommended to add additional information to the part of speech tagger to accurately identify disagreements and increase the diversity and size of the dataset. This would help mitigate the impact of sentence ambiguity on the models' performance.

Regarding the fluctuations observed in the training and validation graphs, it is due to the small data used to train the model. We used 1170 total datasets from these 947 were used for training. By feeding more data, it is possible to minimize the fluctuations since deep learning needs a wide variety of training samples. The BiLSTM model achieved better accuracy than the LSTM model, potentially due to its ability to capture information from both past and future time steps.

In general, the experimental results indicate the effectiveness of the developed models in their respective tasks. The models showed promising accuracy and loss values, with the BiLSTM model outperforming the LSTM model.

6   Conclusion

We developed part of speech tagging and syntax error detection models by undertaking dataset collection, preprocessing, DEAP-based parameter tuning, training, and evaluating the LSTM and BiLSTM algorithms for the Ge'ez language.

To optimize the performances of the proposed models, Ge'ez POS tagging and Ge'ez syntax error detection using deep learning algorithms(LSTM and BiLSTM), different strategies were used. The study used the DEAP hyper-parameter optimization algorithm to automatically search for optimal hyper-parameter configurations, which efficiently explored the hyper-parameter space. Early stopping is implemented during model training to prevent over-fitting and conserve computational resources.

Additionally, a train-validation-split comparison was conducted to evaluate the models' performance using different train-validation splits, such as 70-15-15 and 80-10-10. This comparison helped identify the better train-validation-split ratio.

The results showed the effectiveness of the optimized models, with the LSTM model achieving an accuracy of 94% and 92.31% in the Gz_POS and Gz_SED tasks, respectively, and the BiLSTM model achieving an accuracy of 95.01% and 94.5% in the Gz_POS and Gz_SED tasks. Notably, the BiLSTM model outperformed the LSTM model with some accuracy differences, showing the significance of the optimization techniques.

In general, the use of DEAP, early stopping, and train-validation-split comparison offered a systematic and strong approach to enhance the models' performances in addition to the embedding layer which allows the model to understand the semantics relationships between the words in the sentence.

By developing part of speech tagging and tuning the hyper-parameters of the deep learning algorithms using hyper-parameter optimization algorithms like DEAP, we can handle the syntax errors in Ge'ez. The POS tagging and syntax error detection facilitates the preparation of error-free texts and reliable datasets for Ge'ez language processing tasks.

The successful development of these models opens up opportunities for further research and development of higher-level natural language processing applications in Ge'ez, including syntax error correction, machine translation, semantic analysis, and text summarization. Additionally, this work provides a foundation for future advancements in the field, ultimately contributing to the improvement of language education, translation, analysis, and summarization in the Ge'ez language.

7   Recommendation

Expansion of the Dataset: Although this work has collected a considerable dataset, further expansion of the dataset enhances the performance and generalizability of the models since deep learning algorithms need a large number of datasets to learn from and improve performances.

Error Correction Model: Building upon the syntax error detection model developed in this work, future research can focus on developing an error correction model by incorporating sequence-to-sequence mechanisms and transformers to suggest corrections.

Language Extension: The current work focuses on syntax error detection and POS tagging. Future research can extend the scope by incorporating other linguistic features, such as semantic analysis, named entity recognition, and sentiment analysis to create a wide-ranging NLP toolkit for the Ge'ez language.

Linguistic Resource Development: Since the Ge'ez language is under-resourced, efforts should be made to develop linguistic resources specific to this language. This includes the collection and annotation of corpora that can be used for training and evaluating NLP models. Annotated datasets for various tasks, such as syntax parsing and semantic role labeling, can also be useful for future research.

Comparative Analysis: Conducting comparative analysis with other languages that share similar characteristics with Ge'ez can provide insights into the applicability of the developed models. Researchers can identify transferable techniques and adapt them to improve Ge'ez language processing by studying related languages.

By considering these future steps, researchers can continue to advance the field of Ge'ez language processing, overcome the challenges posed by under-resourced languages, and contribute to the broader domain of natural language processing.

References

1. Abebe, G.: Geez to English Machine Translation Using RNN. Ph.D. thesis, BDU (2021). http://dspace.orghttp//ir.bdu.edu.et/handle/123456789/13149
2. Adamkew, K.: Ge'ez-Amharic Machine Translation Using RNN. Ph.D. thesis, BDU (feb 2021). http://dspace.orghttp//ir.bdu.edu.et/handle/123456789/12380
3. Asmare, H.S., Yibre, A.M.: Geez part of speech tagging using deep learning approaches. In: 2023 International Conference on Information and Communication Technology for Development for Africa (ICT4DA), pp. 181–186 (Oct 2023). https://doi.org/10.1109/ICT4DA59526.2023.10302264
4. Berhanu, B.: Design Ge'ez Factoid Question Answering System (GFQAS) Using Deep Learning Approach. Ph.D. thesis, BDU, Bahir Dar, Ethiopia (jul 2021). http://ir.bdu.edu.et//handle/123456789/12642
5. Dessalegne, Y.A.: Deep Learning Based Amharic Grammar Error Detection. Ph.D. thesis, BDU (2020). https://ir.bdu.edu.et/handle/123456789/12747
6. Desta, B.: Design and Implementation of Automatic Morphological Analyzer for Ge ' ez Verbs. Ph.D. thesis, AAU (2010). aau.etd.edu.et
7. Dillmann, A.: Ethiopic Grammar-2nd-ed-enlarged-and-improved. London: Williams & Norgate, 2nd edn. (1907). https://www.logos.com/product/33588/ethiopic-grammar-2nd-ed-enlarged-and-improved
8. Feleke, F.C.: Morphology based spelling checker for geez language. Hawasa University, Thesis (2023)
9. Fortin, F.A., Rainville, F.M.D., Gardner, M.A., Parizeau, M., Gagné, C.: Deap: evolutionary algorithms made easy. J. Mach. Learn. Res. **13**(70), 2171–2175 (2012). http://jmlr.org/papers/v13/fortin12a.html

10. Gebru, A.T.: Design and Development of Amharic Grammar Checker. Ph.D. thesis, AAU (2013). http://etd.aau.edu.et/bitstream/handle/123456789/917/AynadisTemesgen.pdf?isAllowed=y&sequence=1

11. Gidey, M.A.: Dependency-based Tigrinya Grammar Checker. Ph.D. thesis, AAU (2021)

12. Huang, S., Wang, H.: Bi-lstm neural networks for chinese grammatical error diagnosis. In: Proceedings of the 3rd Workshop on Natural Language Processing Techniques for Educational Applications (NLPTEA2016), pp. 148–154 (2016). https://aclanthology.org/W16-4919

13. Kassahun, A.Y., Fantaye, T.G.: Design and develop a part of speech tagging for ge'ez language using deep learning approach. In: 2022 International Conference on Information and Communication Technology for Development for Africa (ICT4DA), pp. 66–71 (Nov 2022). https://doi.org/10.1109/ICT4DA56482.2022.9971153

14. Katamba, F.: Morphology. Modern linguistics series. Palgrave Macmillan (1993). https://books.google.com/books?id=E-ldR8a1WdEC

15. Lambdin, T.O.: Introduction to Classical Ethiopic (Ge'ez). The president and fellows of Harvard College (1978). https://brill.com/abstract/title/38183

16. Lee, L.H., Lin, B.L., Yu, L.C., Tseng, Y.H.: Chinese grammatical error detection using a CNN-LSTM model. In: Proceedings of the 25th International Conference on Computers in Education, ICCE 2017 - Main Conference Proceedings, pp. 919–921 (2017)

17. Manaye, W.: Designing Geez Next Word Prediction Model Using Statistical Approach. Ph.D. thesis, BDU, Bahir Dar, Ethiopia (2020). http://ir.bdu.edu.et//handle/123456789/11275

18. Memhir Betremariam, A.: Lisane Ge'ez. EthioFagos Book Center, 1st edn (2023)

19. Mercer, S.A.B.: Ethiopic Grammar with Chrestromathy and Glossary. Oxford University Press (1920)

20. Molaligne, T.B.: Amharic Grammar Syntax Error Detection and Correction Model Using Deep Learning. Ph.D. thesis, BDU (2022)

21. Musyafa, A., Gao, Y., Solyman, A., Wu, C., Khan, S.: Automatic correction of Indonesian grammatical errors based on transformer. Appli. Sci. (Switzerland) 12(20) (2022). https://doi.org/10.3390/app122010380

22. Tesfaye, D.: A rule-based afan oromo grammar checker. Inter. J. Adv. Comput. Sci. Appli. 2(8), 126–130 (2011). https://doi.org/10.14569/ijacsa.2011.020823

23. Workie, W.A.: Information Extraction Model from Geez Texts. Ph.D. thesis, BDU (2021). http://dspace.orghttp//ir.bdu.edu.et/handle/123456789/13220

24. Yimam, S.M., Biemann, C.: Current status , issues , and future directions for ethiopian natural language processing (nlp). Tech. rep., Universität Hamburg (2019). https://seyyaw.github.io/files/Yimametal_2019_lt4all.pdf

25. Zhu, J., Shi, X., Zhang, S.: Machine learning-based grammar error detection method in english composition. Sci. Program. 2021 (2021). https://doi.org/10.1155/2021/4213791

Tigrinya End-to-End Speech Recognition: A Hybrid Connectionist Temporal Classification-Attention Approach

Bereket Desbele Ghebregiorgis[✉][ID], Yonatan Yosef Tekle[ID],
Mebrahtu Fisshaye Kidane[ID], Mussie Kaleab Keleta[ID],
Rutta Fissehatsion Ghebraeb[ID], and Daniel Tesfai Gebretatios[ID]

Mai Nefhi College of Engineering and Technology, Mai Nefhi, Maekel Zone, Eritrea
berivaman@gmail.com, yonismo.mc@gmail.com, tatunufti11@gmail.com,
mussiehk216@gmail.com, ruthfisahatsion@gmail.com,
danieltesfai353@gmail.com

Abstract. The latest improvements in end-to-end Automatic Speech Recognition (ASR) systems have achieved outstanding results and have thus enabled the creation of state-of-the-art models for well-resourced languages. However, most languages, such as Tigrinya, are under-resourced, discouraging field efforts. Tigrinya is a Semitic language with over nine million speakers. This paper presents the first hybrid Connectionist Temporal Classification (CTC) with an attention-based end-to-end speaker-independent ASR model for Tigrinya. This initiative constructed new text and speech corpora encompassing multiple domains and thorough pre-processing, which amounted to about 170,000 phrases and sentences of text and 30 h of speech corpus. Data augmentation was applied to generate synthetic data for better generalization capability. A Recurrent Neural Network Language Model (RNN-LM) was also used for post-processing to complement the model to achieve even better results. Multiple experiments were conducted with different settings and parameters. Whilst keeping the data size/split constant and employing various combinations of data augmentation techniques along with varying LM's vocabulary size showed improved performances, increasing the vocabulary size from 5k to 20k resulted in minute decoding improvement. Our best model exhibited a Character Error Rate (CER) of 14.28% and a Word Error Rate (WER) of 36.01%, which is significant considering this end-to-end approach is the first of its kind for the under-resourced Tigrinya language.

Keywords: Tigrinya · Automatic Speech Recognition · CTC with Attention · Data augmentation · Noise removal · Low Resource Language

1 Introduction

Automatic speech recognition (ASR) is a field of study where a computer is tasked to identify and transcribe the sounds produced in human speech. ASR

T. G. Debelee et al. (Eds.): PanAfriConAI 2023, CCIS 2068, pp. 221–236, 2024.
https://doi.org/10.1007/978-3-031-57624-9_12

has shown significant progress over the last decade and has now reached a state where it has become a mainstay in human-computer interaction [17]. A lifetime of work has gone into today's cutting-edge ASR pipelines. The conventional approach to building such systems, notably a Large Vocabulary Continuous Speech Recognition (LVCSR) system, enlists multiple distinct components like phonetic, specialized lexicons, acoustic, and language models [23]. Such approaches typically require linguistic experts and resources such as a tokenization, grapheme-to-phoneme dictionary, and phonological rules [33]. Each of these components is trained in isolation. Consequently, inaccuracies in a single component would significantly impact the subsequent modules. All these factors make engineering and training an ASR system with superior efficiency problematic.

As traditional ASR has several drawbacks, a compelling substitute methodology for training ASR models, i.e., an end-to-end approach, has been proposed recently. The 'end-to-end' strategy replaces most components with a single deep artificial neural network. Thus, it significantly simplifies the development process by eliminating the manual requisite for the compiling, alignment, clustering, and HMM framework commonly utilized to construct cutting-edge ASR models. End-to-end ASR approaches rely on utterances and their corresponding transcriptions and use a single end-to-end ASR algorithm to train the acoustic model. This provides ease and revolutionizes the building of such systems. End-to-end ASR systems convert a series of acoustic inputs into a series of likelihoods for tokens, particularly graphemes and phonetic units. Connectionist temporal classification (CTC) [14], recurrent neural network (RNN) transducer [13], attention-based encoder-decoder [6], and their hybrid models [15,18] are examples of end-to-end architectures for ASR.

The CTC technique guides the training of recurrent neural networks (RNNs) without understanding how input and output streams, which may be of mismatched lengths, correspond. The attention-based model trains the decoder with information from preceding labels, whereas the CTC model can confidently assume connections between labels. Employing a hybrid system is advantageous, particularly for the constrained CTC alignment. Nevertheless, achieving optimal performance with end-to-end ASR systems requires substantial training data. Conversely, many languages need more data due to the high costs of acquiring speech data and its corresponding transcribed text. Out-of-vocabulary (OOV) words and low data resources pose a considerable hurdle in ASR systems. Typically, end-to-end ASR methods rely solely on coupled acoustic and language data. They may experience OOV issues in the absence of additional language data. In this research, we have set out to create a hybrid CTC-Attention-based end-to-end speaker-independent ASR for the Tigrinya Language.

2   The Tigrinya Language

Tigrinya is a member of the Semitic languages, which in turn falls in the Afro-Asiatic family. It has over 9 million speakers and is spoken mainly in Central Eritrea as the official national language and the Tigray region of Ethiopia as a regional language [24]. It was first written in the 13th century in a text on the local laws for the district of 'Logosarda' in southern Eritrea, using a version of the Ge'ez script [28]. It is a morphologically complex language similar to other Semitic languages. The writing system of Tigrinya is based on the old Ge'ez script along with Tigre and Amharic languages [1]. Tigrinya is distinguished from the classical Ge'ez language by using phrasal verbs and a word order that places the main verb last rather than first in the sentence.

2.1 Tigrinya Writing System

The Tigrinya writing system is a subset of what is known as the "Ethiopic" writing system or "Ethiopic syllabary", originally developed for Ge'ez. The Ethiopic script is an abugida: each symbol represents a consonant+vowel syllable, and the symbols are organized in groups of similar symbols based on both the consonant and the vowel. Tigrinya's writing system is a minor variation of the Amharic and Ge'ez writing system and is written from left to right. It makes no distinction between upper and lower-case letters. The graphical unit in this writing system is a consonant followed by a vowel. There are seven vowels in Tigrinya. These are ኧ/ə/, ኡ/u/, ኢ/i/, ኣ/a/, ኤ/e/, እ/ɨ/ and ኦ/o/. Tigrinya characters are made up of thirty-two base symbols, each with seven orders[1] that represent the seven vowels. There are also five derived letters, each with only five orders (vowels) rather than seven orders (where the second and seventh orders are missing). The overall number of graphemes (consonant+vowel) in the Tigrinya language is 249 [12]. The language also has its own punctuation marks and number system.

2.2 Phonology

Although most phones exist in other languages, some are not common in most known languages. These are: t(ጥ), k(ቅ), p(ጵ), ts(ጽ), tʃ(ጭ), ʕ(ዐ), ħ(ሐ), x(ኽ), ɲ(ኝ) and x(ቕ) [1].

2.3 Morphology

The complexity of the Tigrinya language comes from the root pattern morphology it follows. A set of consonants forms the root to make the base of a word, whereas the pattern changes some consonants to vowels, deriving a word with the same meaning but referring to a different gender, number, mood, person, etc. [4,27]. An example is shown in Table 1.

3 Related Works

There is limited research on ASR for the Tigrinya language, as is the case for any other low-resource language. Previous works of speech recognition employed statistical or conventional approaches such as Hidden Markov Models (HMM). For

---

[1] The seven diacritics are used together with the base letters to form unique letters. The diacritics are commonly known as orders.

Table 1. Word Formation.

	Tigrinya	Latin
Root Word	በላዕ/bilie'i/	eat
3rd person singular (Male)	በለዐ/belie'u/	he ate
3rd person singular (Female)	በለዐ/belie'a/	she ate
1st person singular	በለዐ/belie'ə/	I ate

instance, [2] first investigated triphone and consonant-vowel (CV) syllable-based models for the Amharic language using HMM and outlined the CV syllable-based model as the better alternative. Then, HMM-based speech recognition for the Amharic language was built using CV syllables as acoustic modelling units [3]. The work continued by [26] by adding hybrid acoustic models containing phone and context-dependent syllable units to develop a morpheme-based speech recognition. Tachbelie et al. [25] also presented a comparative study of different acoustic modelling units (syllables, triphone, and hybrid) and language modelling units (words and morphemes). As for the Tigrinya language, [11] developed a sub-word based LVCSR for Tigrinya using HMM. Gebregergs [10] also built a DNN-HMM-based speech recognition for Tigrinya, but it is designed to recognize isolated words instead of continuous recognition.

However, the conventional systems mentioned above pose numerous difficulties in building ASR systems. They consist of separate acoustic, pronunciation, and external language models and, therefore, need domain expertise for building and training each module. Furthermore, these systems require extensive data alignment to map each frame to its corresponding label. This makes constructing a large dataset for this conventional approach virtually infeasible. However, later advances in speech recognition have shifted the focus to end-to-end approaches to achieve state-of-the-art results. With the introduction of CTC, there was no more need for alignment or segmentation. This was achieved with the help of the 'blank' label [14]. Chorowski et al. [6] and Hori et al. [15] also designed attention-based models due to the downsides of CTC, although attention had its limitations. To harness the benefits of both methods (CTC and attention), hybrid models have been developed using Multi Task Learning (MTL) [15,18]. However, OOV issues persisted, and [16] used RNN-LM for the language model to alleviate the issue. Many end-to-end based ASR systems exist, albeit primarily for well-resourced languages. In their work, Wang et al. [29] developed an end-to-end Mandarin speech recognition, merging Convolutional Neural Net (CNN) and Bidirectional Long Short-Term Memory (BLSTM). The approach employs CNN for acoustic feature extraction, incorporates BLSTM to capture past and prospective contextual information, and utilizes CTC to decode. The success of this method is attributed to the abundant labeled speech data available to Wang et al. [29].

However, there needs to be more research on building ASR systems using end-to-end approaches for under-resourced languages. Emiru et al. [9] realized this and designed an end-to-end Amharic ASR combining CTC and attention model with phoneme-based Byte-Pair-Encodings (BPE). This model achieved decent results by utilizing data augmentation techniques to add to the small amount of existing Amharic read speech corpora. To the best of our knowledge, no Tigrinya language ASR using an end-to-end approach has been reported.

4   Dataset

Gathering sufficient and quality data is a challenging and costly task. This fact is further prominent in accessing data from under-resourced languages such as Tigrinya. The data construction process encompasses two phases: the text and speech corpus preparation.

Since no comprehensive Tigrinya corpus is widely available, we had to construct our own. The main source of text data was the Eritrean national newspaper of the Tigrinya language, 'Haddas Ertra' ranging from the publication year 2009–2020, which includes multiple domains such as culture, history, politics, law, business, agriculture, health, relationships, social, sport and so on. The text was collected from multiple domains to uphold the diversity of the text corpus. The collected text was then pre-processed using a custom Python program to split paragraphs into sentences, remove punctuation marks, change numbers to their respective Tigrinya word formats, handle abbreviations, etc. This enabled the gathering of clean text data ready for recording. The resulting text corpus contains 170,000 phrases and sentences, of which over 20,000 were used for recording. Moreover, a vocabulary size of 20,000 words was used for training the language model.

On average, a single speaker would read about 400 sentences/phrases, amounting to 30 min in duration. The read speech corpus was collected using an Android application called 'Lig-Aikuma' [5], which can quickly gather speech data on different mobile phones and tablets. The recording was collected with the following specifications: mono channel, sampling frequency at 16 kHz, and bitrate at 256 kb. The collected speech data was then pre-processed using a 'denoiser' tool [8]. It is a pre-trained speech enhancement model that removes any stationary or non-stationary background noise in the audio. Further manual pre-processing was also done to clean damaged, clipped, or corrupt audio files. The resulting speech corpus includes 30 h of recording gathered from a total of 51 Tigrinya native volunteer speakers, out of which 26 were male & 25 were female.

Finally, the data split was done randomly. The training, validation, and test data split comprises of non-overlapping speakers (i.e., no speaker utterance is used in more than one data split category). This guarantees the usage of utterly new unlearnt utterances and new speakers in any data split category, making the split suitable for baseline experimentation. Additionally, due to the small size of the corpus, data augmentation techniques were used. These techniques enable better model generalization by garnering synthetic data from the collected dataset [20]. It complements the collected data to increase the quantity of training data by alleviating the over-fitting problem and creating robust models [19].

This becomes particularly useful for building end-to-end ASR for low-resource languages as end-to-end models require large training data sizes. Therefore, we adopted a state-of-the-art data augmentation technique named 'SpecAugment' [21]. The augmentation policies utilized here include time masking, time warping, and frequency masking. Table 2 shows the statistical summary of the collected data.

Table 2. Statistical analysis of the collected audio data (duration is in seconds).

	Entire Dataset	Training split (83%)	Validation split (11%)	Testing split (6%)
Min. Duration	0.99	0.99	1.65	1.79
Max. Duration	18.94	18.94	15.97	11.70
Mean Duration	5.21	5.24	5.12	5.00
Median Duration	4.75	4.76	4.53	4.60
Data Size (in hrs.)	30:05hr	25:05hr	3:13hr	1:47hr
Male Speaker Count	26	17	6	3
Female Speaker Count	25	24	0	1

5   Methods

5.1   CTC

Traditional models (GMM/HMM and Deep Learning HMM) demand that the linguistic units and the audio signal be aligned to be trained. These alignments can be difficult to obtain manually, especially for large datasets. CTC [14] was introduced to provide a method of training RNNs to label unsegmented sequences directly. Given a set of acoustic features $X = [x_1, x_2, \ldots, x_T]$, and the desired output sequence $Y = [y_1, y_2, \ldots, y_U]$. The exact alignment of X to Y is unknown, and the ratio of the lengths of X and Y is also variable. To do this, it creates a pair (X, Y) in which the output is assigned to each input (x_t). As a null delimiter, a blank token is introduced. These predictions are collapsed, and the token is deleted, permitting repeating sequences and periods of "silence." Therefore, it is omitted from decoding and/or loss computation. Its purpose is to align input and output directly without forcing a classification on the vocabulary. An example of CTC alignment can be shown in Fig. 1.

A probability distribution over all potential Ys is then generated. This distribution is useful for estimating a particular output, Y. As a result of summing all possible alignments between the input and the output, the conditional probability, $P(Y|X)$, is calculated. Dynamic programming is used to improve the computation of the CTC loss function. CTC loss function is given by:

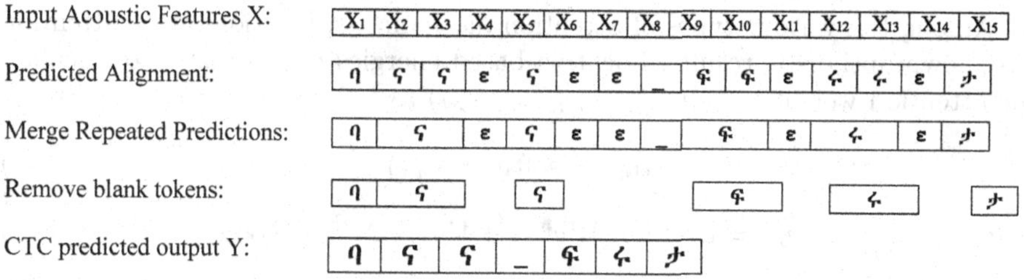

for an input X and output Y = [በ, ና, ና, ,ፍ, ሩ, ታ].

Fig. 1. CTC alignment.

$$\mathrm{L_{CTC}}(X, Y) = -\log \sum_{a \in A_{X,Y}} \prod_{t=1}^{T} \mathrm{P}(a_t \mid X) \tag{1}$$

5.2 Attention-Based Models

In contrast with the CTC method, the attention model instantly predicts each target without requiring any intermediary representation or assumptions. This results in a higher CER than CTC when no external language model is utilized. The encoder-decoder subnetworks train a neural network to transcribe spoken utterances. The encoder converts a speech's acoustic feature sequences to the sequence representation's length (T). With the attention-based approach, the decoder converts high-level features (H) the shared encoder produced into a p output label sequence. The decoder estimates the likelihood of the label sequence of each input feature (F) using the chain rule [13] based on the conditional probability of the label p_u given the input feature H and the preceding labels $p_{1:u-1}$.

$$\Pr(p|F) = \prod_{u} \Pr(p_u|H, p_{1:u-1}) \tag{2}$$

At every step u, the decoder establishes a context vector c_u by considering both the input features H and the attention weight $\alpha_u \sum t$. This contextual vector is key in decoding and capturing the input sequence's relevant information. The attention weight $\alpha_u \sum t$ guides the decoder in focusing on specific elements of the input features, ensuring an effective creation of the context vector at each decoding step.

$$c_u = \sum_{t} \alpha_{u,t} H_t \tag{3}$$

Among the various types of attention mechanisms, location-aware attention has shown optimistic results. Location-based energies $e_{u,t}$ are used to calculate the attention weight $\alpha_u = (\alpha_{u,1}, \alpha_{u,2}, \ldots, \alpha_{u,t})$ as follows:

$$\alpha_{u,t} = \text{softmax}(e_{u,t}) \tag{4}$$

$$e_{u,t} = \omega^T \tanh(W q_{u-1} + V h_t + M f_{u,t} + \alpha) \tag{5}$$

$$f_u = F * \alpha_{u-1} \tag{6}$$

As stated in [9], "ω, V, W, α, and M are trainable parameters, and q_{u-1} is the state of the RNN decoder. $*$ denotes the one-dimensional convolution along the frame axis, t, with the convolution parameter F, to produce the features $f_u = (f_{u,1}, f_{u,2}, \ldots, f_{u,t})$. We can predict the RNN hidden state q_u and the next output p_u with the context vector c_u in Eqs. 7 and 8, respectively."

$$q_u = \text{LSTM}(q_{u-1}, p_{u-1}, c_u) \tag{7}$$

$$p_u = \text{FullyConnected}(q_u, c_u) \tag{8}$$

Here, the LSTM layer in Eq. 7 is unidirectional, and Eq. 8 implements a fully connected feed-forward layer. "sos" and "eos" in the decoder module for attention-based end-to-end speech recognition signify the start and end of a sequence, respectively. When "eos" is emitted, the decoder halts, generating new output labels.

5.3 Multitask Learning (MTL)

Multitask learning techniques were developed as a solution for the attention and CTC disadvantages [31]. Attention typically performs better in end-to-end scenarios; however, it frequently struggles to converge and suffers in noisy environments. Conversely, CTC tends to produce worse quality results due to the conditional independence assumption but is more stable. As a result of the trade-offs between CTC and attention, their combination in multitask learning is extremely valuable. Kim et al. [18] and Xiao et al. [32] were trained to optimize an attention-based encoder-decoder model with CTC and attention. Their training loss is a multi-objective loss (MOL), which is defined as:

$$L_{MOL} = \lambda \log \text{P}_{\text{CTC}}(C|X) + (1 - \lambda) \log \text{P}_{\text{att}}(C|X) \tag{9}$$

where C is the probable character sequence, X is the speech input sequence, and λ is the weight assigned to each loss function with a value ranging from 0 to 1. The CTC objective is denoted by P_{ctc}, and the attention objective is denoted by P_{att}^*. The general framework of MTL is shown in Fig. 2.

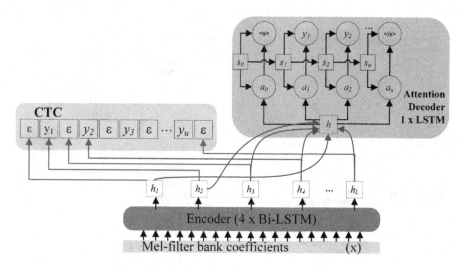

Fig. 2. Hybrid CTC-attention-based end-to-end framework.

6   Experiments

6.1   Experimental Setup

The experiments were carried out using a virtual machine (VM) instance on the Google Cloud Platform (GCP) compute engine. The VM features 16 vCPUs of Intel Skylake, 64 GB RAM, 300 GB SSD storage, NVIDIA TESLA T4 GPU, and runs on the Google Deep Learning OS. This configuration significantly enhanced training speeds, thereby reducing the overall training time. The ESP-net toolkit [30] was utilized for all training purposes, with PyTorch selected as the backend. Data preparation, involving the extraction of acoustic features from Tigrinya speech corpus utterances, was carried out using the Kaldi [22] tool. Mel-Frequency Cepstral Coefficients (MFCC) of 80-dimensional log filter bank feature vectors were extracted. MFCC was used because it captures the main characteristics of speech and has higher performance compared to other methods, such as Linear Predictive Coding (LPC) and Perceptual Linear Prediction (PLP) [7].

The training parameters were configured by consulting multiple research papers (especially for Semitic languages). The optimizer Ada-delta with a batch size of 30 was utilized for 20 epochs of training (maxlen-in = 800 and maxlen-out = 150). The encoder included 4 BLSTM layers with 320 cells, while the decoder included 1 LSTM layer with 300 cells. Sortagrad parameter was set to -1 to feed the samples from shortest to longest. The attention method used was location-based, and the language model contained 1 RNN layer with 1000 units. A patience parameter value of 3 and validation accuracy as an early stop criterion were used to prevent overfitting. The category of the end-to-end model was defined by the CTC-weight (λ) and was represented under multi-task learning

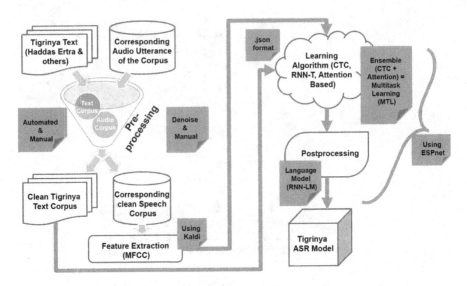

Fig. 3. System Architecture.

as mtlalpha. The value of mtlalpha (both for training and decoding) was fixed at 0.5 for the hybrid CTC-attention end-to-end model [31]. The type of end-to-end model was determined by CTC-weight (λ) and was described in terms of multi-task learning (mtlalpha). The value of mtlalpha (both for training and decoding) was set as 0.5 for the hybrid CTC-attention end-to-end model. As for the LM, various word-level RNN-LMs were prepared using varying sizes of only transcription of audio data. The RNN-LMs consisted of a single LSTM layer with 1000 cells and were trained using Stochastic Gradient Descent (SGD) for 20 epochs. A beam width of 20 and RNNLM-weight of 1.0 were also used for decoding. Figure 3 depicts the overall pipeline used to construct the model.

7 Results and Discussions

Before any deductions are made, it is imperative to consider the effect of the small data size used. It would also be helpful to point out that the experiments were conducted in a manner where the effects of data augmentation were observed before increasing the LM's vocabulary size.

The first experiment conducted showed promising CER results with a fairly appropriate data split (83% training, 11% validation, and 6% testing). However, the result obtained wasn't ground-breaking. Nonetheless, that became a "close to accurate" baseline model selected for comparison purposes throughout our experimentation. The second experiment kept all parameters constant except for adding a data augmentation method called 'SpecAugment' [21] with a combination of maximum time warping, frequency mask, and time mask of 5, 30, and 40, respectively. The effect of integrating a data augmentation technique

Table 3. Experimental Results.

Exp	Model	Vocabulary Size	OOV (%)	CER (%)	WER (%)	Accuracy (%)
1	MTL+Sortagrad	5K	23.7	16.87	43.41	84.75
2	MTL+Sortagrad + SpecAug	5K	23.7	14.47	36.39	88.17
3	MTL+Sortagrad	10K	17.3	17.40	44.60	84.41
4	MTL+Sortagrad + SpecAug	10K	17.3	14.36	36.28	88.32
5	MTL+Sortagrad	20K	11.8	16.75	43.36	84.98
6	MTL+Sortagrad + SpecAug	20K	11.8	**14.28**	**36.01**	**88.45**

Table 4. Experimental Results for the "Amharic character-based end-to-end model in different vocabulary sizes".[*]

Language Unit	Vocabulary Size	Language Model	Acoustic Model	CER (%)	WER (%)
character	10K	WordRNNLM	CTC-attention	26.91%	37.60%
character	20K	WordRNNLM	CTC-attention	25.21%	36.80%

[*] This result is borrowed from the Table 1 in [9]

virtually increased the data size. This showed significant improvement in both CER & WER throughout the experiments.

With such a small data size, the problems of overfitting and OOV words frequently arise. As such, SpecAugment and validation accuracy as early stop criteria were used to prevent overfitting while LM was made use of to alleviate OOV. The SpecAugment, as described in [21], is beneficial in improving generalization, thereby resolving the overfitting problem. The OOV words issue was also moderately addressed by incorporating an RNN-LM. It can be observed that the increase in vocabulary size showed a decrease of half in comparison with the base model's OOV. However, it didn't quite provide a tangible CER/WER or Accuracy enhancement.

It is unclear whether this minute improvement is due to the small vocabulary size or due to the use of word-LMs instead of character-based or even a sub-word-based LM, so further investigation will be necessary. Towards the end, the last experiment showed even further improvement in both CER & WER, resulting in our personal best of 14.28% and 36.01%, respectively. Table 3 shows the overall experimental outcomes, with Figs. 4, 5, and 6 highlighting the best-case CER, Loss, and Accuracy results, respectively. Contrary to the norm, it can be observed that the validation results for CER and Accuracy are higher than their training counterparts. Likewise, validation loss results are also lower than the training loss results. Nonetheless, the results do not imply overfitting and are valid. This is because the SpecAugument is only applied during the training phase, making the prediction more difficult than that in the inference phase.

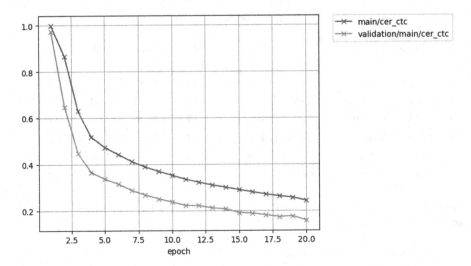

Fig. 4. Exp.6 CER Results.

Furthermore, since no comparable preceding work had been done on the Tigrinya language, it's only logical to present our findings as a base model for upcoming research. Nevertheless, we still decided to compare it to an Amharic end-to-end model constructed using an analogous approach [9]. These two languages are of Semitic origin and share quite a lot, including nearly identical sets of characters, phonemes, and writing systems. The results of the character-based acoustic model incorporating a word-RNNLM in [9] were only applicable for comparison purposes since all the experiments conducted in this paper were of that type. The results they achieved are shown in Table 4.

The best-case of their character-based joint CTC-attention end-to-end model without SpecAugment and with 20K vocabulary size of word-based RNN-LM was a CER of 25.21% and a WER of 36.80%. Moreover, a similar model but with a 10K vocabulary size was a CER of 26.91% and a WER of 37.60%. In contrast with the best-case results of an analogous setup having 20K vocabulary size we obtained, which were a CER of 16.75% and a WER of 43.36%, we have had significantly better results with regards to CER (an improvement of 8.46% and 9.51% in both 20K and 10K vocabulary size). As for the WER, ours was 6.56% and 7.00% more erroneous than theirs. It's important to note that this comparison should be approached cautiously due to variations in languages and data sizes between the works.

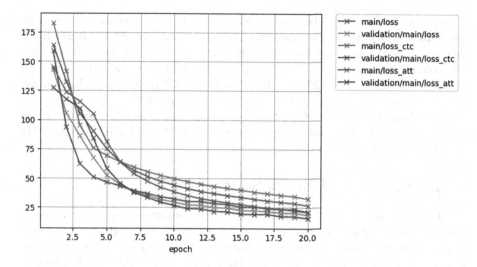

Fig. 5. Exp.6 Loss Results.

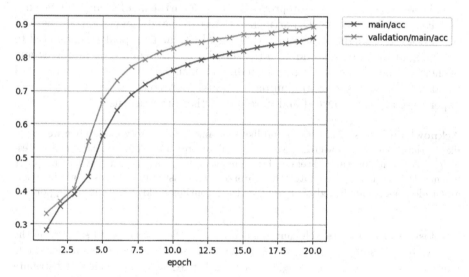

Fig. 6. Exp.6 Accuracy Results.

8   Conclusions

This project aimed to create the first end-to-end Tigrinya ASR system. We made a new multi-domain text and a multi-speaker speech corpus that is suitable for ASR and NLP experimentation in the Tigrinya language. The constructed data was then used to train a hybrid CTC-Attention acoustic model and an RNN word LM. We conducted multiple experiments by varying the LM size and parameters used.

We found that using data augmentation greatly complemented the existing data to outperform the experiments without augmentation. Although the increase in vocabulary size decreased the OOV rates, it did not have any discernable impact on the CER or WER results, pending further research. Our best model exhibited a CER of 14.28% and a WER of 36.01%, which is significant considering this research is the first of its kind for the under-resourced Tigrinya language. Thus, whilst this can be used as a baseline for upcoming studies, results also indicate that the corpus is suitable for further investigation and development of ASR systems.

9   Limitations and Ethics Statement

As discussed above, the major challenge with building ASR for an under-resourced language is the construction of a new corpus. Although we constructed a novel text and speech corpus, it is relatively small. Moreover, the speech collected only includes 'read speech' and a highlander accent, which is predominant throughout Eritrea.

The work presented was approved by the Department of Computer Science and Engineering and the Department of Research and Innovation at the Mai Nefhi College of Engineering and Technology, Eritrea. The speakers consented to recording of their speech, and then the speech was anonymized by removing any of their distinguishing attributes, including their names. The authors annotated the data only to ensure optimum data quality. Moreover, the tools used in this paper were all open-sourced and were put to their intended use.

Acknowledgments. First, we would like to thank the Almighty God. Then we would like to thank Dr. Yonas Meressi, Minister of Transport and Communications Mr. Tesfaslasie Berhane and the EriTel co., Dr. Yemane Keleta, the Department of Computer Science & Engineering, volunteer data donors, and last but not least, our heartfelt gratitude goes to our friends & family for their continuous love and moral support.

Disclosure of Interests. The authors declare that the research data supporting the findings of this study are available from the corresponding author upon reasonable request. The authors retain the right to be the sole party able to provide and distribute the data used in this study.

References

1. The Tigrinya Language. (2021). https://www.ucl.ac.uk/atlas/tigrinya/language.html
2. Abate, S.T.: Automatic speech recognition for Amharic. Ph.D. thesis, Staats-und Universitätsbibliothek Hamburg Carl von Ossietzky (2006)
3. Abate, S.T., Menzel, W.: Syllable-based speech recognition for Amharic. In: Proceedings of the 2007 Workshop on Computational Approaches to Semitic Languages: Common Issues and Resources, pp. 33–40. Association for Computational Linguistics, Prague (2007). https://aclanthology.org/W07-0805

4. Abate, S.T., Tachbelie, M.Y., Schultz, T.: Multilingual acoustic and language modeling for ethio-semitic languages. In: Interspeech (2020)
5. Blachon, D., Gauthier, E., Besacier, L., Kouarata, G.N., Adda-Decker, M., Rialland, A.: Parallel speech collection for under-resourced language studies using the lig-aikuma mobile device app. Procedia Comput. Sci. **81**, 61–66 (2016). https://doi.org/10.1016/j.procs.2016.04.030
6. Chorowski, J., Bahdanau, D., Serdyuk, D., Cho, K., Bengio, Y.: Attention-based models for speech recognition. arXiv:1506.07503 cs.CL (2015)
7. Dave, N.: Feature extraction methods LPC, PLP and MFCC in speech recognition. Int. J. Adv. Res. Eng. Technol. **1**(6), 1–4 (2013)
8. Defossez, A., Synnaeve, G., Adi, Y.: Real time speech enhancement in the waveform domain. arXiv:2006.12847 eess.AS (2020)
9. Emiru, E.D., Xiong, S., Li, Y., Fesseha, A., Diallo, M.: Improving amharic speech recognition system using connectionist temporal classification with attention model and phoneme-based byte-pair-encodings. Information **12**(2), 62 (2021). https://doi.org/10.3390/info12020062
10. Gebregergs, G.: DNN-HMM Based Isolated-Word Tigrigna Speech Recognition System. Master's thesis, Addis Ababa Institute of Technology (2018)
11. Gebretsadik, T.: Sub-word Based Tigrinya Speech Recognizer an Experiment Using Hidden Markov Model, pp. 1–7. GRIN Verlag, Munich (2013)
12. Girmasien, Y.: Qalat Tigrinya ab Srah/Tigrinya Words in Action, 1st edn., pp. 22–30. Brhan Media Services (2011)
13. Graves, A.: Sequence transduction with recurrent neural networks. arXiv:1211.3711 cs.NE (2012)
14. Graves, A., Fernández, S., Gomez, F.J., Schmidhuber, J.: Connectionist temporal classification: labelling unsegmented sequence data with recurrent neural networks. In: Cohen, W.W., Moore, A.W. (eds.) Machine Learning, Proceedings of the Twenty-Third International Conference (ICML 2006), Pittsburgh, 25–29 June 2006. ACM International Conference Proceeding Series, vol. 148, pp. 369–376. ACM (2006). https://doi.org/10.1145/1143844.1143891
15. Hori, T., Watanabe, S., Hershey, J.: Joint CTC/attention decoding for end-to-end speech recognition. In: Proceedings of the 55th Annual Meeting of the Association for Computational Linguistics (Volume 1: Long Papers), pp. 518–529 (2017)
16. Hori, T., Watanabe, S., Zhang, Y., Chan, W.: Advances in joint CTC-attention based end-to-end speech recognition with a deep CNN encoder and RNN-LM. arXiv:1706.02737 cs.CL (2017)
17. Kamath, U., Liu, J., Whitaker, J.: Deep Learning for NLP and Speech Recognition. Springer, Cham (2019). https://doi.org/10.1007/978-3-030-14596-5
18. Kim, S., Hori, T., Watanabe, S.: Joint CTC/attention based end-to-end speech recognition using multi-task learning. In: 2017 IEEE International Conference on Acoustics, Speech and Signal Processing (ICASSP), pp. 4835–4839 (2017)
19. Ko, T., Peddinti, V., Povey, D., Khudanpur, S.: Audio augmentation for speech recognition. In: INTERSPEECH, pp. 3586–3589 (2015)
20. Ma, E.: Data augmentation for audio (2021). https://medium.com/makcedward/data-augmentation-for-audio-76912b01fdf6
21. Park, D., et al.: Specaugment: a simple data augmentation method for automatic speech recognition. In: INTERSPEECH, pp. 2613–2617 (2019). https://doi.org/10.21437/Interspeech.2019-2680
22. Povey, D., et al.: The kaldi speech recognition toolkit. In: IEEE 2011 Workshop on Automatic Speech Recognition and Understanding. No. IEEE Catalog: CFP11SRW-USB. IEEE Signal Processing Society (2011)

23. Sen, S., Dutta, A., Dey, N.: Audio Processing and Speech Recognition: Concepts, Techniques and Research Overviews. Springer, Singapore (2019)
24. T., D.P., William, B.: Ethiopic Writing. The World's Writing Systems. Oxford University Press (1996)
25. Tachbelie, M.Y., Abate, S.T., Besacier, L.: Using different acoustic, lexical and language modeling units for ASR of an under-resourced language - amharic. Speech Commun. **56**, 181–194 (2014). https://doi.org/10.1016/j.specom.2013.01.008
26. Tachbelie, M.Y., Abate, S.T., Besacier, L., Rossato, S.: Syllable-based and hybrid acoustic models for amharic speech recognition. In: Third Workshop on Spoken Language Technologies for Under-resourced Languages, SLTU 2012, Cape Town, 7–9 May 2012, pp. 5–10. ISCA (2012). https://www.isca-speech.org/archive/sltu_2012/tachbelie12_sltu.html
27. Tedla, Y.K., Yamamoto, K., Marasinghe, A.: Tigrinya part-of-speech tagging with morphological patterns and the new nagaoka tigrinya corpus. Int. J. Comput. Appl. **146**(14) (2016)
28. Voigt, R.: Tigrinya. In: Weninger, S. (ed.) The Semitic Languages: An International Handbook, Handbücher zur Sprach- und Kommunikationswissenschaft/Handbooks of Linguistics and Communication Science, vol. 36, pp. 1153–1169. De Gruyter Mouton, Berlin (2011)
29. Wang, D., Wang, X., Lv, S.: End-to-End Mandarin Speech Recognition Combining CNN and BLSTM. Symmetry **11**(5), 644 (2019). https://doi.org/10.3390/sym11050644
30. Watanabe, S., et al.: Espnet: end-to-end speech processing toolkit. arXiv preprint arXiv:1804.00015 (2018)
31. Watanabe, S., Hori, T., Kim, S., Hershey, J.R., Hayashi, T.: Hybrid CTC/attention architecture for end-to-end speech recognition. IEEE J. Select. Topics Signal Process. **11**(8), 1240–1253 (2017). https://doi.org/10.1109/JSTSP.2017.2763455
32. Xiao, Z., Ou, Z., Chu, W., Lin, H.: Hybrid CTC-attention based end-to-end speech recognition using subword units. In: 2018 11th International Symposium on Chinese Spoken Language Processing (ISCSLP), pp. 146–150. IEEE (2018)
33. Yu, D., Deng, L.: Automatic Speech Recognition, pp. 13–48. Springer, London (2016)

DNN-Based Supervised Spontaneous Court Hearing Transcription for Amharic

Martha Yifiru Tachbelie[2], Solomon Teferra Abate[2], Rosa Tsegaye Aga[1],
Rahel Mekonnen[1(✉)], Hiwot Mulugeta[1], Abel Mulat[1], Ashenafi Mulat[1],
Solomon Merkebu[1], Taye Girma Debelee[1,3], and Worku Gachena[1]

[1] Ethiopian Artificial Intelligence Institute, Addis Ababa, Ethiopia
`rosa.tsegaye@aii.et`, `rahelmekonen9@gmail.com`, `hiwot.mulugeta05@gmail.com`,
`abelmulat21@gmail.com`, `ashenafimulat0239@gmail.com`, `solmer2014@gmail.com`,
`{taye.girma,worku.gachena}@aii.et`
[2] School of Information Science, Addis Ababa University, Addis Ababa, Ethiopia
`{martha.yifiru,solomon.teferra}@aau.edu.et`
[3] Department of Electrical and Computer Engineering, Addis Ababa Science
and Technology University, Addis Ababa, Ethiopia

Abstract. Research in the area of Automatic Speech Recognition (ASR)
for Ethiopian languages, especially for Amharic has been conducted since
2001. However, the ASR systems have not been used for real-world applications such as court hearing transcription. In this paper, we present our
endeavour towards the development of a DNN-based supervised spontaneous court hearing transcription system for Amharic. Speech and text
corpora that are required for the development of the ASR system are collected from the Ethiopian Federal Courts. The text data has been cleaned
through a series of pre-processing tasks while the speech data has been
manually segmented and aligned with the corresponding transcription. A
Deep Neural Network (DNN) based ASR system has been developed using
90% of the pre-processed data and the remaining 10% has been used as
an evaluation set. The back-end of the system interface has been developed using Laravel PHP while the front-end has been developed using Java
script, HTML5, and CSS. The database is developed with MySql and for
the API python programming language has been used. The performance
of the court transcription system has been evaluated in terms of word error
rate (WER) of the ASR system on the evaluation set. When the ASR system is manually evaluated on 6000 segments (from the evaluation set) by a
transcriber who listened to each of the audio segments, it achieved a WER
of 7.6%. Based on the result, we can conclude that this transcription system can be used by the Federal courts of Ethiopia with minor editing of
the transcription output.

Keywords: Court hearing transcription · Spontaneous speech
recognition · Amharic · Deep Neural Network

© The Author(s), under exclusive license to Springer Nature Switzerland AG 2024
T. G. Debelee et al. (Eds.): PanAfriConAI 2023, CCIS 2068, pp. 237–249, 2024.
https://doi.org/10.1007/978-3-031-57624-9_13

1   Introduction

There are over 7000 languages in the world but only a few of the languages are supported by speech processing applications. Although there are different factors that contribute to the absence of speech-processing applications for most of the world's languages, lack of language and technological resources stand on the first line of the obstacles.

Despite the efforts made by different tech companies such as Meta and Google to include less-resourced languages by developing multilingual language resources as well as models and providing a limited free cloud-based computing infrastructure, the development of speech applications that solve societal problems is hardly achieved. This is true in Ethiopia where there are more than 80 less-resourced languages.

Nowadays, there are efforts in the preparation of speech corpora for Ethiopian languages by different researchers and institutions, including the international ones. To mention some attempts, a 20 h of speech corpus was prepared by [1]. Moreover, corpora of four Ethiopian languages (Amharic, Afan Oromo, Tigrigna, and Wolaytta) have been prepared by the Addis Ababa University under the thematic research funding framework [2]. A 100 h of Amharic speech corpus, which includes the 20 h of Amharic speech corpus that has been prepared by [1], is also prepared by [7]. Recently, the Ethiopian Artificial Intelligence Institute (EAII) is also aggressively working towards the preparation of speech corpora for Ethiopian languages such as Amharic, Afan Oromo, Tigirigna and Somali [6]. However, almost all the speech and language resources are from broadcast media and religious domains. To the best of the authors' knowledge, there are no domain-specific speech and language resources, including the judicial domain. With regard to computational resources, the Ethiopian Artificial Intelligence Institute has high-performance computational resources that enable the research and development of speech applications.

Research in the area of speech applications, in general, and automatic speech recognition (ASR), in particular, for Ethiopian languages started in 2001 when [4] developed a Consonant-Vowel syllable recognition system for Amharic. Since then, several researches in the area have been and being conducted for different Ethiopian languages. However, the potential benefit of automatic speech recognition systems in solving real societal problems has not yet been fully explored. To our knowledge, there are only limited attempts towards the development of speech-enabled applications [3,6,15,16].

On the other hand, there is a pressing need for speech-processing applications in various sectors in Ethiopia. For example, in the legal sector, court hearing recordings are transcribed manually by human transcribers. As it is known manual transcription is time taking. Literature shows that an average transcriber should work for up to five hours to transcribe an audio of one hour long. But in most cases depending on different factors such as recording quality, background noise, and variation in the speaker's background (accent, different mother tongue, and emotion), the speed of transcription goes very low and the time may go up to nine hours to transcribe one-hour long audio.

From the onsite visit and discussions that had been done at the Ethiopian Federal courts, transcribers need around eight hours to transcribe forty-five minutes long audio recordings. This indicates that court hearing transcription is one of the time-consuming tasks. Consequently, there is a huge backlog of the transcription task that results in a long and unnecessary delay to the court proceedings. Moreover, the task of manual court transcription is not an easy task and is full of responsibilities. During transcription, a word mistake may result in a misjudgment. But the manual transcription is prone to error due to its tedious nature. In addition, it results in several health problems on the transcribers such as problems of ear, eye, kidney, leg, and back.

ASR engines are being used as the front-line solution for such tasks in the legal sector. Literature shows that court reporters and agencies are employing ASR for accurate and effective legal transcription.

In this paper, the development of an Amharic court hearing transcription system has been presented. The contribution of the paper is twofold: One is the development of the court hearing transcription system that fills the existing gap in exploring the potential of automatic speech processing applications towards solving real societal problems in less-resourced and technologically less-favored languages, and the second contribution is the preparation of an in-domain speech corpus for a less-resourced language, Amharic.

The rest of the paper is organized as follows. The next Sect. 2 provides a brief review of works in Amharic ASR and speech-enabled applications. Data collection and pre-processing will be presented in Sect. 3. Sections 4 and 5 present the development of the ASR model, which is the heart of the automatic court hearings transcription system, and the development of the automatic court hearings transcription system. The evaluation of the transcription quality is presented in Sect. 6. Finally challenges, conclusions and future works are presented in Sects. 7 and 8, respectively.

2   Related Works

There are a large number of works that have been done in the area of ASR for Ethiopian languages in general, and Amharic in particular. However, since our concern in this paper is on the application of the ASR system, we will review only works that have been conducted with the aim of using ASR or speech technology for different purposes. The first work that used ASR for command and control purposes is [15]. In this work, a word-based ASR system was developed using Hidden Markov Model (HMM) and used to command and control Microsoft Word using Amharic speech. Since the number of words used in the ASR system is too small and since word-based models are used, the accuracy was very high, close to 100%.

Another work which is conducted in the judicial domain was [3]. The system is named "Brana" "ብራና", after the writing material made from animal skin which is used in ancient times. Brana is a dictation system intended for real time transcription. Two HMM-based Amharic ASR systems were developed using one and half hours of spontaneous speech from the judicial domain and twenty hours of Amharic read speech corpus prepared by [1]. The best-performing spontaneous and read speech ASR systems, as reported by the authors, have WERs of 54.86% and 16.13%, respectively. The high WER of the spontaneous ASR system is attributed to the nature and amount of the speech data used.

Another work is an Amharic dictation system which was developed in a project funded by the Ministry of Communication and Information Technology (MCIT) [16]. In this work, a morpheme-based ASR system was developed using HMM and the twenty hours of read speech corpus [1]. The ASR system was then integrated with Microsoft's word-processing software. The WER of the ASR system was 17.2% when it was tested offline. After the integration of the ASR system to Microsoft Word, the dictation system was tested using four speakers and resulted in an average WER of 21.8%.

Recently, the Ethiopian Artificial Intelligence Institute released a free mobile application for public use that is called Learn and Transcribe ET Lang [6]. The application transcribes Amharic and Afan Oromo speeches either pre-recorded or live. In addition, the application can be used to learn the basics of Afan Oromo, Tigirigna, and Somali languages for Amharic speakers. This application has been developed using more than 1000 h of audio data (for each Amharic and Afan Oromo languages), that have been prepared at the institute.

Although there are many Artificial intelligence (AI)-based automatic court transcription service providers, especially for English and there are many works in the automatic court transcription for different languages, a review of only a few of the works, that are published, will be presented.

[8] developed an automatic transcription system for a judicial domain in Polish and Italian languages. They have developed the ASR system for both languages using HMM. In the development of the ASR systems, they have used out-of-domain data and adapted using a certain amount of in-domain speech (none for Italian and 24 h for Polish) and text data. The ASR performance was measured on audio data acquired in the courtrooms of Naples and Wroclaw for Italian and Polish, respectively. The WERs of the ASR systems were around 40%, for Italian, and between 30% and 50% for Polish.

Another work is an automatic Slovak dictation system for judicial domain [12]. The ASR system is developed using HMM and 250 h of speech data (120 h recorded from the judicial domain and 130 h of out-of-domain speech data). The speech data from the judicial domain is recorded in sound studio condition and the other in an office and conference room environment. The text corpus used for the language modeling purpose consists of 120 million sentences or more than two billion words. The best model has a WER of 5.26%.

A more recent work is an automatic speech transcription system for UK supreme court hearings [13]. In this work, the authors investigated domain adaptation of off-the-shelf ASR systems. They used two off-the-shelf ASR systems that are Amazon Web Services and OpenAI's Whisper model. They have tried to reduce the WERs that occur when the ASR systems are used in the legal domain by using an in-domain language model and by infusing common phrases. This way, 9% and 8% WER improvements have been obtained over the Amazon Web Service and the Whisper model, respectively. The lowest average WER that they have achieved is 11.6%.

As can be observed from the above reviews of local works, almost all the works are done using HMM. Moreover, most of them were developed using a small amount of data and not used in solving real societal problems. The current work fills the gap by developing an ASR system using the state-of-the-art machine learning algorithm, DNN, using a relatively large amount of in-domain speech and text corpora, and using the ASR system for solving a real societal problem which is court hearing transcription.

3   Data Preparation

Since there is no in-domain text and speech data for the development of an ASR system for court hearing transcription in any of the Ethiopian languages, in general, and Amharic, in particular, one of the main tasks conducted in this work is data preparation. The ASR system developed in this work consists of three models that are language model (LM), Lexical model (LeM), and Acoustic model (AM). To develop these models text-only and speech-with-text data have been prepared. In this section, a brief description of the collection and pre-processing of the two types of data will be presented in Subsects. 3.1 and 3.2, respectively.

3.1   Data Collection

The first task in data preparation is the collection of the raw data that has been generated as a result of the day-to-day activities of the court system. The data collection/acquisition department of the Ethiopian Artificial Intelligence Institute collected the required data in collaboration with domain experts and higher officials of the Ethiopian Federal courts. The following subsections present the collected data.

Text Data. For the development of the language and lexical models of the ASR system, the required text corpus has been collected from the three Federal Courts of Ethiopia that are the Supreme, High, and First Instance Courts. From the different archives of the court systems, a large amount of text-only data that is not tokenized at the sentence level has been collected. These text data has been tokenized into sentences using the common sentence-ending markers (። ፤ ? ፦ ! ፣). However, since there were long sentences consisting of more than 100 words, further tokenization has been done using keywords that usually indicate sentence ending such as ነው meaning "it is" or "be" and ነበር meaning "it was" or "used to". After the tokenization, it was possible to get around 6.4 million sentences that are not pre-processed.

Speech Data. The speech data that is required for the development of the acoustic model needs to consist of the corresponding text transcription. Therefore, the recorded speech along with the transcription has been collected. The transcription of the audio data has been produced as the day-to-day activities of the manual transcription in the court systems. With the active collaboration of the Federal courts of Ethiopia and domain experts from the courts, it was possible to collect a total of 3,590 h of audio data with the corresponding transcription.

3.2   Data Pre-processing

Text Data. A text data that is to be used for the development of language models should pass through a series of text pre-processing tasks. Thus, the collected in-domain text data should be pre-processed to be usable in the development of an automatic court hearing transcription system. The common pre-processing tasks that are conducted in the ASR community and that have been conducted in this work include removing special characters and punctuation marks, expansion of abbreviations, expansion of numbers, normalization of redundant characters, and removing sentences that consist of characters and words in other languages.

The collected text consists of different special characters such as emojis, and hidden encoding representations that are inserted in the process of manual typing or document conversion among text editors and/or operating systems. As per the investigation that has been done in this study, some of them do not affect the meaning and syntax of the text. They are, therefore, removed from the text. When the special characters affect the meaning and syntax of the sentences, the sentences that consist of such characters are removed from the text. Similarly, punctuation marks have also been removed.

The use of abbreviations is the culture of court hearing transcription that has been reflected in the collected text. The abbreviations have been automatically extracted from the collected text. They are then manually expanded and represented as a dictionary of abbreviations that is used to automatically expand all the abbreviations in the text.

In the same manner, the numbers in the text have been extracted and categorized into different categories such as date, account numbers, and telephone numbers. Based on their categories, they have been expanded to their textual representations automatically.

In the Amharic writing system, there are a number of characters that represent the same sound and are used interchangeably without affecting the meaning of the words. For example, the ፕ and ሐ in ፕብት and ሐብት represent the same sound and the meaning of both words is wealth. Similarly, the አ and ዐ in አለም and ዐለም represent the same sound and both words mean world. If such characters are not normalized, they will result in several variations of the same word that are considered different. Therefore, all such redundant characters have been normalized without affecting the size of the text rather than increasing the occurrence of the words that consist of such characters.

In the collected text, a large number of non-Amharic characters and words have been used for different purposes. Since it was difficult to translate or transcribe them into Amharic, all the sentences that contain these characters and words have been removed. This and the above pre-processing tasks reduced the size of the collected text data to only 3.8 million sentences.

Speech Data. Since the collected speech data consists of very long audio recordings that cannot be used for the development of the acoustic model, they have to be segmented into short segments along with their corresponding transcriptions. This task was a very expensive task that could not be done automatically. Therefore, it has been done manually engaging several experienced transcribers from the Federal courts of Ethiopia for several months. The task is time-consuming, labor-intensive, costly, and tedious. Due to time and financial constraints and the difficulty of the task, it was not possible to segment and transcribe all the collected speech data. Although 3,590 h of transcribed speech data have been collected, only 151,028 segments have been prepared for the development of the acoustic model. The total length of these transcribed speech segments is 159.56 h. Compared to the amount of speech data used for the development of court hearing transcription systems for other languages, for instance, [12], the amount of speech data used in the current work is small.

4   ASR System Development

Automatic speech recognition systems can be developed in two main approaches that are Supervised and Unsupervised. The unsupervised approach is introduced as a solution for the problem of data scarcity. For most of the ASR application areas, the supervised approach outperforms. With the aim of choosing the best method, experiments have been conducted using an unsupervised transfer learning approach. It has been found out that the ASR system developed this way has a performance which is inferior to the one developed using the supervised approach. Supervised ASR system can also be developed in two methods. The first is the classical approach that uses acoustic, language and lexical models. The second and recent approach is the end-to-end approach which could not yet outperform the classical one in large application domains. Thus, in this study, the classical approach has been adopted for developing the ASR system that is used in the court hearing transcription system. The development of the lexical, language and acoustic models is presented in the following subsections.

4.1   Lexical Model

In the classical approach of ASR development, the sequence of phonemes for each word in the transcription to be aligned with the speech data is generated by the lexical model. In this work, two types of lexical models have been prepared. These are the training lexical model which is used during acoustic model training and the decoding lexical model which is used during decoding or recognition. The list of word types for the development of the training lexical model has been extracted from the transcription of the training speech data. For the decoding lexical model, all the word types extracted from the pre-processed text data and the training transcription have been used. Following the approach used in previous works for Amharic, the lexical models, both the training and decoding, have been developed automatically. In the approach, an Amharic character has been represented as a sequence of a consonant and a vowel. The size of the training lexical model is 101,022 word types while the size of the decoding lexicon is 1,423,546. The total number of phones that have been used to phonetically transcribe is 38.

4.2 Language Model

For the development of the language model (LM), the pre-processed text-only data and the text transcription of the training speech have been used. The total amount of text data that is used for language modeling purpose is 3.9 million sentences or 1,423,544 words. Since the n-gram modeling approach has been adopted, the well-known language model development toolkit that is called SRILM [14] has been used. The model is smoothed using the unmodified Kneser-Ney smoothing techniques [5].

4.3 Acoustic Model

Out of the total audio segments with transcription prepared for the acoustic model development purposes, 90% have been used for training and the rest for evaluation purposes. For the development of the acoustic model(AM), a total of 135,880 utterances that is 143.63 h of speech have been used.

Since the hybrid HMM-DNN approach has been used, the Hidden Markov model-Gaussian Mixture Model (HMM-GMM) based model has been developed. The HMM-GMM acoustic model has been trained using the well-known Kaldi ASR toolkit [9]. A context-dependent HMM-GMM-based AM has been built using 39-dimensional mel-frequency cepstral coefficients (MFCCs), and to each of which cepstral, mean, and variance normalization (CMVN) have been applied. A fully continuous 3-state left-to-right HMM architecture has been used. Linear Discriminant Analysis (LDA) and Maximum Likelihood Linear Transform (MLLT) feature transformation have also been applied for each of the models followed by Speaker Adaptive Training (SAT) using an offline transform, feature space Maximum Likelihood Linear Regression (fMLLR).

To train the DNN-based acoustic model, the best HMM-GMM model has been used to get alignments. The same training speech data (a total of 135,880 segments that is 143.63 h of speech) has been used for the extraction of 40-dimensional MFCCs without derivatives, 3-dimensional pitch features, and 100-dimensional ivectors for speaker adaptation. The Factored Time Delay Neural Networks architecture with additional Convolutional layers (CNN-TDNNf) has been used. The DNN architecture has 15 hidden layers (6 CNN followed by 9 TDNNf) and a rank reduction layer.

5   The Transcription System

The model that has been developed is integrated with the transcription system and has become ready for the end users. The end users of this transcription system are the transcribers who work at the Ethiopian federal courts. The system back end has been developed using Laravel PHP, the front end using JavaScript, html5, and CSS, and the database using MySQL. The ASR model has been integrated with the system with an Application Programming Interface (API) that has been developed using Python programming language. Figure 1 shows the architecture of the transcription system.

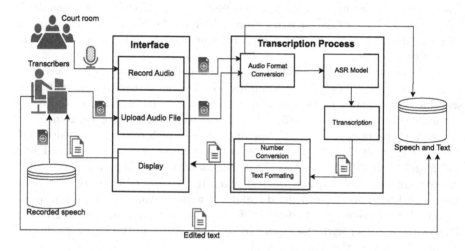

Fig. 1. Court Hearing Audio Transcription System Architecture

The system has multiple features such as the administrator side for the user account, database management, and the transcriber side to feed the audio for transcription. The system gives the transcribed text in a few minutes depending on the length of the audio. The other two major features of the system are audio data preparation and text-to-number conversion. Since the ASR system accepts and transcribes waveform audio with a sampling rate of 16 kHz and bit rate of 16, the input audio files need to be converted to comply with these requirements. The audio data preparation, therefore, includes the audio format conversion such as from mp3 to wave, sampling rate conversion to 16 kHz, and bit rate conversion to 16 bits. Thus, the transcription system accepts audio data of any format, except audio data that is recorded with less than 16 kHz and 16 bits. The text-to-number conversion converts the textual forms of numbers into digits in the transcription output of the ASR system. In addition, the system works by following the Ethiopian Federal Court rules and regulations; for example, the transcribed text can not be edited after a given time without the supervisor and/or judge's permission.

6   Evaluation

As it has been presented in the previous sections, the court hearing transcription system has different components that work together in the process of transcribing input audio to the corresponding text transcription. These components have been evaluated using different formats and different lengths of audio files and are found to be properly working with any audio format (except audio data that are recorded with less than 16 kHz and 16 bits) and length.

The three components of the ASR system (the language, lexical, and acoustic models) have been evaluated. The language and the lexical models have been evaluated in terms of perplexity and Out-of-Vocabulary (OOV) rate, respectively. Together with these models, the performance of the acoustic model is evaluated in terms of Word Error Rate (WER). The WER can be based on human judgment or can be based on a reference transcription prepared for the evaluation purpose.

In the current work, the OOV rate of the decoding lexical model which consists of about 1.4 million words is computed on the transcription of the evaluation set which consists of 15,148 sentences and 114,329 unique words. The OOV rate of the model is found to be 1.85%. This is due to the morphological complexity of the Amharic language and the spelling errors, which could not be corrected. The perplexity of the language model is also computed on the same test text and found to be 703.25. The perplexity shows that there is a significant difference between the training and the test texts. In the manual transcription, there are misspelled words due to the difficulty of writing Amharic characters using keyboards that are primarily designed for languages written in Latin alphabets. This factor together with the morphological complexity of the language affected the performance of the language model.

The WER of the ASR system has been evaluated based on 10% of the manually segmented and transcribed speech data. The evaluation set is 15.93 h long. The WER of the system compared to the test transcription is 30.8%. A detailed error analysis has been done, and it has been found out that the reference transcription has a lot of errors. That means the WER rate does not reflect the actual performance of the ASR system. Therefore, the performance of the ASR system has been evaluated manually by experienced transcribers who listened to 6,000 utterances and cross-checked with the output transcription of the ASR system. When evaluated this way, the WER became 7.6%, which means an accuracy of 92.4%.

To confirm the ASR model that has been developed and used in the court hearing transcription system has a better performance compared with the large ASR models that are openly available, a transcription of speech data from the court domain has been done using the openly available large models. For this purpose, Open AI's Whisper [11] and Meta's Massively Multilingual Speech (MMS) [10] have been considered. Through readings, we have learned that Whisper is a multilingual and multitask ASR model that is not fine-tuned to any specific task. As a result, it does not beat task or domain-specific models such as ours. Thus, the test is conducted using Meta's MMS model. MMS is a multilingual model that supports ASR and text-to-speech for more than 1100 languages, including Amharic. The model was trained using speech data that are available online and are mostly from religious domain. When it is tested using Amharic speech, Meta's MMS model usually does a good job with short sentences. However, when it is given a spontaneous speech from the court domain, it didn't perform well. For example, when it was evaluated with a spontaneous speech data that consists of 27 words from the court domain, Meta's MMS wrongly recognized 18 words (WER of 66.67%) while our ASR model made only two errors (WER of 7.41%) as shown in Table 1. The errors committed by our ASR model are in the recognition of person names.

Table 1. Comparison of DNN-based Domain Specific and Meta's MMS ASR Models.

	ASR models	
	DNN-based	MMS
Total no. of words in the output	27	23
No. of substitution errors	2	12
No. of insertion errors	–	1
No. of deletion errors	–	5
Total no. of errors	2	18
WER	7.41	66.67

Regarding the speed, the ASR model by itself delivers the transcribed text in a few seconds and/or minutes depending on the size of the audio. For example, for two hours of audio, it takes around two minutes to deliver the transcription. But the court transcription system has a couple of features such as audio format and number conversions that may add more seconds and/or minutes for the transcription output to be presented for the user.

7   Challenges

The main challenges have been faced in the data preparation. Since the text-only as well as the speech with transcription data were not in the format required for model training, and they are bulk data, the data preparation required much effort and took most of the time, financial and computational resources.

One of the pre-processing tasks that have been done is the segmentation of long audio files with their transcription. A tool that helps the transcribers listen and segment the audio and the corresponding transcription at the sentence level has been developed. Despite the support of the tool, during audio segmentation, the transcribers segmented or split a word which results in meaningless sounds. Some transcribers also did not write the words in the proper way because of their accents and backgrounds. As a result, the segmented audio and transcription should be audited and edited multiple times that made the pre-processing task costly and time-consuming.

The other challenge is on the text data preparation for the language model training. The text data that has been collected from the Ethiopian Federal Courts contains a large number of misspelled words, concatenated words, and unknown characters. Since the size of the data, in terms of the number of sentences, is in millions, it was challenging to have a clean text by fixing each and every error manually.

8   Conclusion and the Way Forwards

This paper presented an endeavour towards the development of a DNN-based supervised spontaneous court hearings transcription system for Amharic. Despite the challenges in data preparation and morphological complexity of the language, notable results have been achieved. The ASR system, which is the backbone of the court transcription system, consists of three models that are lexical, language, and acoustic models. These models have been trained using large sizes of audio with transcription and text-only data that have been collected from the Ethiopian Federal Courts. The accuracy of the ASR system has reached 92.4% in transcribing court hearings. The ASR system performance is the court transcription system's performance as well.

The following works are planned to be done in order to enhance the system's capabilities and make the model robust. These are increasing the amount of the data set so as to improve the performance of the three components of the ASR system, using speaker diarization techniques, and incorporating real-time transcription features. Ultimately, the aim is to make the court hearings transcription system a valuable tool for the Ethiopian legal sector, offering efficient, accurate, and accessible transcription solutions.

Acknowledgement. We would like to acknowledge the Ethiopian Federal Courts and the domain experts who are working closely with the transcription system development team of the EAII. We are also grateful to all the transcribers who were involved in the speech segmentation and transcription task.

References

1. Abate, S.T., Menzel, W., Tafila, B.: An amharic speech corpus for large vocabulary continuous speech recognition. In: INTERSPEECH 2005 - Eurospeech, 9th European Conference on Speech Communication and Technology, Lisbon, Portugal, September 4–8, 2005, pp. 1601–1604. ISCA (2005). https://doi.org/10.21437/INTERSPEECH.2005-467

2. Abate, S.T., et al.: Large vocabulary read speech corpora for four Ethiopian languages: Amharic, Tigrigna, Oromo and Wolaytta. In: Calzolari, N., et al. (eds.) Proceedings of the Twelfth Language Resources and Evaluation Conference, pp. 4167–4171. European Language Resources Association, Marseille, France (2020). https://aclanthology.org/2020.lrec-1.513

3. Alemayehu, B.A., Abate, S.T.: Brana(ብራና): application of amharic speech recognition system for dictation in judicial domain. In: Proceedings of the Ethiopian Information Technology Anual Conference (2015)
4. Berhanu, S.: Isolated Amharic consonant-vowel (CV) syllable recognition : an experiment using Hidden Markov Model (HMM). MSc. thesis, School of Information Studies for Africa, Addis Ababa University (2001)
5. Chen, S.F., Goodman, J.: An empirical study of smoothing techniques for language modeling. Comput. Speech Lang. **13**(4), 359–394 (1999)
6. Ethiopian Artificial Intelligence Institute. https://www.aii.et/projects/. Accessed 31 Oct 2023
7. Gebreegziabher, N.H., Nürnberger, A.: An amharic syllable-based speech corpus for continuous speech recognition. In: Martín-Vide, C., Purver, M., Pollak, S. (eds.) SLSP 2019. LNCS (LNAI), vol. 11816, pp. 177–187. Springer, Cham (2019). https://doi.org/10.1007/978-3-030-31372-2_15
8. Lööf, J., Falavigna, D., Schlüter, R., Giuliani, D., Gretter, R., Ney, H.: Evaluation of automatic transcription systems for the judicial domain. In: 2010 IEEE Spoken Language Technology Workshop, pp. 206–211 (2010). https://doi.org/10.1109/SLT.2010.5700852
9. Povey, D., et al.: The kaldi speech recognition toolkit. In: Automatic Speech Recognition and Understanding - ASRU (2011). https://api.semanticscholar.org/CorpusID:1774023
10. Pratap, V., et al.: Scaling speech technology to 1,000+ languages. arXiv preprint arXiv:2305.13516 (2023)
11. Radford, A., Kim, J.W., Xu, T., Brockman, G., McLeavey, C., Sutskever, I.: Robust speech recognition via large-scale weak supervision. In: International Conference on Machine Learning (2022)
12. Rusko, M., et al.: Slovak automatic dictation system for judicial domain. In: Vetulani, Z., Mariani, J. (eds.) LTC 2011. LNCS (LNAI), vol. 8387, pp. 16–27. Springer, Cham (2014). https://doi.org/10.1007/978-3-319-08958-4_2
13. Saadany, H., Breslin, C., Orasan, C., Walker, S.: Better transcription of UK supreme court hearings. arXiv preprint arXiv:2211.17094 (2022)
14. Stolcke, A.: Srilm - an extensible language modeling toolkit. In: Interspeech (2002). https://api.semanticscholar.org/CorpusID:1988103
15. Tachbelie, M.Y.: Application of Amharic speech recognition system to command and control computer: an experiment with microsoft word. MSc. thesis, School of Information Studies for Africa, Addis Ababa University (2003)
16. Tachbelie, M.Y., Abate, S.T., Abebe, E.: Development of Morpheme Based Dictation System with the Support of Automatically Constructed Dictionary. School of Information Science, Addis Ababa University, Tech. rep. (2015)

Typewritten OCR Model for Ethiopic Characters

Bereket Siraw Deneke[1], Rosa Tsegaye Aga[1]([✉]), Mesay Samuel[2],
Abel Mulat[1], Ashenafi Mulat[1], Abel Abebe[1], Rahel Mekonnen[1],
Hiwot Mulugeta[1], Taye Girma Debelee[1,3], and Worku Gachena[1]

[1] Ethiopian Artificial Intelligence Institute, Addis Ababa, Ethiopia
rosa.tsegaye@aii.et
[2] Arbaminch University, Arbaminch, Ethiopia
mesay.samuel@amu.edu.et
[3] Department of Electrical and Computer Engineering, Addis Ababa Science and
Technology University, Addis Ababa, Ethiopia

Abstract. Optical Character Recognition (OCR) is the electronic conversion of images of computer-written, typewritten, handwritten, or
printed text into machine-encoded text from a scanned document and a
photo of a document. In Ethiopia, documents such as historical, office,
and official documents have been documented in handwritten and typewritten form, until recently. Thus, a large number of historical and essential documents are still in hard-copy form and at risk of disaster to be
lost. Computer-written and handwritten OCR have been developed for
different language characters including Ethiopian languages (Ethiopic
characters), But not typewriter-written OCR for Ethiopic scripts. Like
handwritten documents, large historical documents are typewritten documents in Ethiopia. Thus, the typewritten OCR is mandatory to preserve these documents. This study focuses on building an OCR model
for typewritten documents that are written on Ethiopic characters. For
the study, different Ethiopic characters have been collected from typewritten documents, and 290 distinct characters have been segmented to
construct augmented data to form various character variations and simulate the complexities encountered in real-world typewritten Amharic
texts and enhance the adaptability of the OCR model. This technique
aims to approximate the diversity inherent in the data. The model training framework leverages the capabilities of Tesseract, an open-source
OCR engine, in conjunction with the artificially generated training set.
The Tesseract's existing Amharic OCR model has been deployed as a
base model, and the fine-tuning process has been adopted in a layered
approach by employing 45,000 samples and spanning 4,800 iterations.
The model has been evaluated using character error rate (CER). As per
the evaluation, the model performed with 13% CER on the test set. For
this study, the Tesseract model before fine-tuning and the Google Lense
platform has been used as a baseline to evaluate the performance of the
model. Accordingly, our model has outperformed both baselines by more
than 10% margin.

T. G. Debelee et al. (Eds.): PanAfriConAI 2023, CCIS 2068, pp. 250–261, 2024.
https://doi.org/10.1007/978-3-031-57624-9_14

Keywords: CNN · RNN · OCR · Tesseract · Transcribe

1   Introduction

In the rapidly evolving technological landscape, harnessing innovation to address challenges is pivotal for the present and future of Africa. Just as the inception of writing laid a solid foundation for human progress throughout history, Optical Character Recognition (OCR) stands as a contemporary pillar of state-of-the-art technology in our era.

A significant portion of Ethiopian archives comprises handwritten and typewritten documents [1]. This paper presents an improved OCR model tailored for recognizing typewritten Amharic text, thereby facilitating the digitization and preservation of this invaluable corpus. The importance of this OCR extends beyond the ease of information storage and retrieval; it simplifies the process of converting text into Braille. Such an application of OCR plays an undeniable role in promoting inclusively by granting visually impaired individuals access to a corpus that they could not access. According to the latest Ethiopian census, there are well over half a million visually impaired individuals, including students, lawyers, teachers, researchers, and artists [2]. This underscores the critical need for an enhanced OCR model to create a world where visually impaired people access a wealth of knowledge. This is just one example of how an improved OCR model is crucial, not only in developed countries but also in developing countries like Ethiopia.

While there are various OCR models publicly available, both free and paid, developing a customized OCR model tailored for typewritten Ethiopic script languages is a significant step forward. It allows for the localization of state-of-the-art technologies and offers greater flexibility for model improvement as needed [1,3]. In this paper, we conduct a comparative analysis among different well-known OCR models and evaluate the Character Error Rate (CER) of our new model in comparison to the others. This showcases the effectiveness of our tailored approach, which takes full advantage of Tesseract's capabilities while adapting them to the unique demands of Ethiopic script.

Our newly created model harnesses the power of the Tesseract OCR engine, widely regarded as one of the best OCR engines due to its flexibility and accuracy. Tesseract stands out for several reasons. First, it's an open-source OCR engine, which means it's continuously improved and maintained by a global community of developers [4]. This ensures that it remains up-to-date and adaptable to various languages and scripts, including Ethiopic scripts.

As we delve into the task of adapting an OCR model for typewritten Ethiopic script recognition, specific challenges have been encountered. One of the main challenges is insufficient training data. The development of effective OCR models heavily relies on access to substantial volumes of training data. To prevent data hallucination, it is also essential for the data to have balanced character repetition frequencies. In the case of Ethiopic scripts, there is limited availability of labeled data, which necessitates the use of creative approaches for training set creation and model adaptation. In this case, the preparation of box files for each image file is time-consuming, as it requires manual character-by-character selection.

This study introduces a pioneering OCR model tailored for typewritten Ethiopic characters, addressing a longstanding void in artificial intelligence language technology for overlooked languages namely Amharic. By creatively utilizing segmented character facets and simulating diverse datasets, the accuracy of transcription has been enhanced. This achievement underscores the potential and applicability of AI-driven OCR technology for local languages, contributing to a more comprehensive and inclusive technological landscape with limited resources and time.

The rest of the paper is organized as follows. Section 2 provides a brief review of works in OCR models and applications that have been done in local and international languages. The methodology that contains Data preparation and the Tesseract framework is explained in Sect. 3. The experiments on the model training and the state-of-the-art OCR platforms are explained in Sect. 4. Results and Conclusion are explained in Sects. 5 and 6, respectively.

2   Related Work

In the field of OCR, several notable advancements and solutions have been developed, some of which are particularly relevant to this study on typewritten Amharic script recognition. This section provides an overview of key related works and their contributions.

i2OCR is an online OCR service that has gained recognition for its versatility in recognizing text from images [5]. While it supports a variety of languages, including Amharic, its adaptability to typewritten Amharic script and its accuracy in recognizing the unique characteristics of this script have been somewhat limited [6].

Google Lens, a part of the Google ecosystem, offers OCR capabilities for various languages, including Amharic. It utilizes machine learning and computer vision techniques to extract text from images [7]. While Google Lens is capable of recognizing typewritten Amharic text, its accuracy in this context is not as high, particularly when it is compared to our model.

EasyOCRConverter is another tool that offers OCR services for multiple languages, and it is acknowledged for its user-friendly interface [9,18]. However, when it is applied to the accurate transcription of typewritten Amharic corpus, it exhibits a relatively high CER that reaches as much as 67%.

Moreover, the scientific investigation with regard to Ethiopic typewritten character recognition is hardly addressed. There are however several notable attempts in Ethiopic handwritten and computer written document recognition setting a significant foundation to other related works including the typewritten document recognition. [10–12] are the earliest works for printed and handwritten Amharic character recognition employing traditional template matching and structural analysis methods. Following the advancements in machine learning, deep learning techniques have been applied for Ethiopic printed [13–15] and handwritten [16,17] OCR. [18] followed the same deep learning approach to recognize historical Ethiopic handwritten texts. Due to the dataset size limitation, most of these works have adopted various data augmentation techniques to cope up with the deep learning training.

While these existing solutions offer valuable insights and capabilities, our work seeks to address the specific challenges posed by typewritten Amharic script recognition. We aim to contribute to this body of study by introducing a tailored OCR model that takes into account the unique characteristics and limitations of Amharic script, with a focus on enhancing accuracy and accessibility for visually impaired individuals.

Our approach not only involves model development but also innovative data preparation techniques to overcome the scarcity of labeled data, a challenge that is particularly prominent in the context of Amharic script recognition. This paper aims to provide a comprehensive assessment of our model's effectiveness and its potential impact on AI-driven OCR technology for local languages, including Amharic, in the evolving technological landscape.

3   Methodology

3.1   Data Preparation

The Ethiopian writing system adheres to a left-to-right and top-to-bottom manner of writing. It comprises 35 distinct characters that are organized into seven groups, and each group represents a syllable that consists of a consonant followed by a vowel. Additionally, it incorporates 54 symbols for labialization, 20 symbols for numerals, and nine signs for punctuation marks. Consequently, this script encompasses a total of 328 unique characters. Figure 1 shows the Amharic typewritten text sample data.

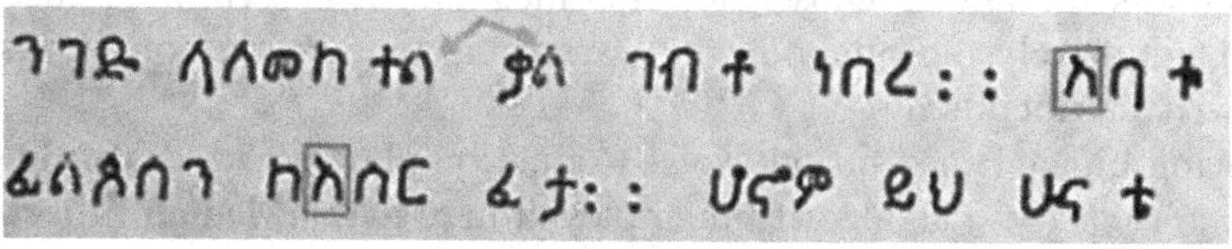

Fig. 1. Sample Amharic typewritten text

The unique characteristics of the Amharic alphabet with its intricate combination of consonants and vowels pose distinctive challenges for OCR model training and recognition. Additionally, the subtle differences between most non-basic characters can be challenging to discern, and this distinction can further fade due to ink leakage in typewritten prints as shown in Fig. 1. In Fig. 1, the red boxes show two different Amharic characters from the same basic character group but with distinct sounds and representations. However, in the typewritten print, it can be challenging to notice these differences. This is just one example; there are other cases where characters may appear similar but are different.

Typewritten Amharic texts often exhibit variations in color tones due to ink inconsistencies, which can obscure the natural appearance of characters. This variability can pose difficulties in character recognition and necessitate robust image reprocessing techniques. This can also be observed in Fig. 1. The green arrows indicate two identical Amharic characters represented slightly differently. The top part of the leftmost character is missing, due to ink inconsistencies that caused some parts of the character to be invisible. This, on its own, might not pose much of a challenge because there is no Amharic character that resembles it, even with the upper part of the character removed. However, the problem arises during the preprocessing step of transcription. During transcription, various image preprocessing techniques are applied. During these preprocessing stepts, these partially lost parts are completely removed as shown in Fig. 2. This results in an incorrect transcription outcome.

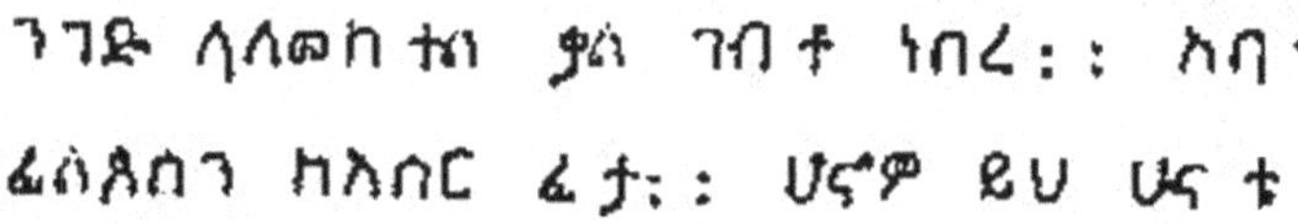

ንገዱ ሰሰወስ ትሰ ያሰ ገበ ቸ ከበረ፡፡ አበ ቶ

ፈሰጸበን ከአሰር ፈታ፡፡ ህናም ይህ ህና ቸ

Fig. 2. The result of Image Binarization and Noise Reduction of the sample image

The preprocessing steps adopted in this study encompass a suite of techniques crucial for preparing typewritten Amharic text images for effective OCR model training. The first step, Image Binarization, is implemented using the Adaptive Thresholding technique, particularly Otsu's method. This method automatically calculates the optimal threshold separating foreground and background pixels by maximizing inter-class variance, effectively converting the image into a binary format. Otsu's method is chosen for its efficiency in determining the optimal threshold adaptively, accommodating variations in lighting conditions within the typewritten documents.

Following Image Binarization, Noise Reduction plays a pivotal role in enhancing character clarity. The algorithm employed here is the Median Filtering technique. Its selection is based on its ability to effectively remove impulse noise or speckles while preserving edge information, crucial for maintaining character integrity in the image. Median Filtering computes the median value within a specified neighborhood, making it robust against outliers commonly found in typewritten documents.

Background Removal, a subsequent step, relies on Morphological Operations like erosion and dilation. These operations are preferred due to their ability to effectively eliminate background elements surrounding characters. Erosion systematically erases pixels from the edges, aiding in separating characters from the background, while dilation expands pixels, further isolating characters. The combination of erosion and dilation via Morphological Operations ensures improved segmentation accuracy, vital for isolating characters of interest.

One of the foremost challenges in training an effective OCR model lies in the acquisition of a substantial amount of labeled training data. To address this challenge, we employ the utilization of a simulated dataset. This simulated dataset is crafted through a series of steps, each aimed at emulating real-world conditions and intricacies encountered in typewritten Amharic text recognition. Thus, the simulated dataset generation process begins with the extraction of segmented characters from typewritten scanned images, encompassing all Amharic character representations. These have been extracted characters undergo rigorous preprocessing, wherein background elements are effectively removed employing adaptive thresholding techniques. The training data preparation is shown in Fig. 3.

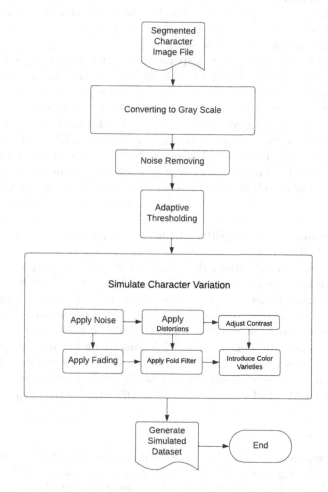

Fig. 3. Training data preparation

The incorporation of various character variations in the training dataset plays a crucial role in simulating the intricacies present in authentic typewritten Amharic texts, a vital aspect for developing a robust Optical Character Recognition (OCR) model. These diverse forms of character variations, illustrated in Fig. 4, encompass several essential elements. Firstly, the introduction of random noise imitates the inherent imperfections often found in typewritten documents due to environmental factors or machinery inconsistencies. Controlled distortions further emulate real-world scenarios where characters might be slightly altered or warped, ensuring the OCR model learns to decipher such deviations accurately.

Fluctuations in contrast within the dataset expose the model to varying levels of brightness and darkness, mirroring the diverse conditions under which typewritten texts may be scanned or captured. The subtle fading effects incorporated contribute to replicating the gradual degradation of text over time, a common occurrence in aged documents. Additionally, the application of fold filter effects introduces creases and folds, replicating the challenges posed by physical damage or irregularities in the scanned documents.

Furthermore, the inclusion of a spectrum of color tones designed to mimic aged or oil-stained papers contributes significantly to the model's adaptability. Understanding and decoding characters under different color shades is essential for accurate recognition, especially in the case of older or deteriorated documents.

This comprehensive approach to dataset augmentation serves dual purposes. Firstly, it mitigates the limitations posed by a scarcity of labeled training data, effectively expanding the model's learning potential by presenting a more extensive range of potential inputs. Secondly, it significantly enhances the adaptability and robustness of the OCR model to effectively handle the complexities inherent in real-world typewritten Amharic texts. By exposing the model to a wide array of variations, it learns to generalize better and perform more accurately when faced with diverse typewritten documents, ultimately improving its overall performance and usability.

The total augmented dataset that has been used for training and validation counts to 45,000 samples. From this, 40%, 20%, 25%, and 15% are from syllables, labialized symbols, numerals, and punctuation marks respectively. This distribution takes into account the intrinsic importance and likely occurrence frequency of each character group in Amharic text. Allocating a higher percentage to syllables acknowledges their central role while ensuring a balanced representation across other character categories. Accordingly, 90% of the samples have been used for training and the remaining 10% has been used for validation. For the test set, a separate set of newly generated 45,000 samples has been used.

3.2 Tesseract

In this section, the strategies that have been used for the study are elaborated, emphasizing the importance of customization and fine-tuning of the Tesseract OCR engine to meet the specific demands of typewritten Amharic text recognition.

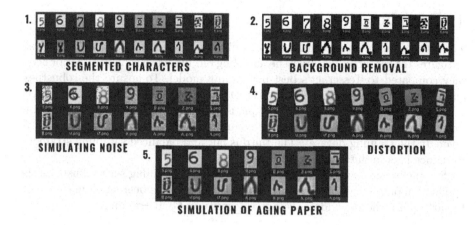

Fig. 4. Preprocessed Segmented characters

The choice to incorporate a pre-trained Tesseract model into our OCR methodology stems from several pivotal considerations. Tesseract, a highly acclaimed open-source OCR application, is renowned for its adaptability and precision in character recognition. Its internal architecture, which integrates a hybrid of Convolutional Neural Networks (CNN) and Recurrent Neural Networks (RNN), aligns seamlessly with the intricacies of typewritten Amharic text. Despite the existence of alternative models leveraging CNNs and/or RNNs, the decision to fine-tune Tesseract presents a pragmatic advantage. The open-source nature of Tesseract not only ensures continuous refinement but also facilitates adaptability to diverse languages, particularly catering to the nuances of Ethiopic scripts. By leveraging the robust and evolving capabilities of Tesseract, our OCR model is positioned to benefit from a well-established foundation, contributing meaningfully to the existing research landscape while aligning with the recognized strengths of this esteemed OCR engine.

The process of training a model using Tesseract involves several crucial steps. It requires a substantial amount of training data specific to the target language, in this case, Amharic. This data has been used to train the model on the unique characteristics of Amharic script, including its characters, ligatures, and diacritics. The training process involves iterative adjustments to improve accuracy continually.

Additionally, the model can be fine-tuned to recognize typewritten Amharic text more effectively. This involves adjusting hyperparameters and optimizing the model for the specific requirements of the task. Fine-tuning ensures that the OCR model can handle the nuances of typewritten Amharic text with the highest degree of accuracy.

4 Experiment

The integration of these simulated datasets into our training process effectively mitigates the resource scarcity issue, enabling our model to achieve superior accuracy compared to Tesseract's best-performing model. To ensure the robustness and effectiveness of our model, various methods have been applied to address the over-fitting and bias. The first technique is regularization; for this, dropout and weight decay have been applied. The second technique is data augmentation to diversify the training dataset. The third is fair and balanced sampling strategies to reduce bias in dataset representation.

Sample images have been also included from the training set to illustrate the application of our approach. Additionally, this study has been applying state-of-the-art approaches to address resource scarcity in OCR training.

4.1 Model Training

Different hyperparameters have been considered to train the model. These are iteration, sample size, existing open-sources base model and learning rate. During the training, the model performance was evaluated using character error rate(CER). The hyperparameter values have been trained with a given range of values. For the iteration, the value range between 1000 and 19000 has been used. For the sample size, the value range between 467 and 54000 has been used. For the learning rate, the value between 0.01 and 0.0001 has been considered.

4.2 State of the Art OCR Platforms

For the proposed model's performance evaluation, the state-of-the-art platforms have been used as a baseline. Even if there are no existing OCR platforms that are available for typewritten images, the existing and widely used platforms, Google Lens and Tesseract have been used. Google Lens has been selected as a benchmark for our OCR model comparison due to its widespread recognition and accessibility within the Google ecosystem, making it highly relevant for a broad user base. Google Lens offers OCR capabilities for multiple languages including Amharic, and provides a valuable reference point for typewritten Amharic script recognition. Its presence in the technology landscape allows for a pragmatic assessment of our model's performance against a widely used solution and provides insights into potential enhancements and innovations required for OCR technology, particularly in underrepresented languages such as Amharic. The other platform that has been used as a baseline is the Tesseract base model. In this study, a comparative analysis has been running with Google Lens, and Tesseract's optimized model tailored for the Amharic corpus to gauge its relative accuracy and effectiveness.

5 Result

As explained in the experiment section, multiple experiments have been run to find an outperforming model from the different hyperparameter values. As a result, the model trained on the value of iteration 4800, learning rate 0.0001, and sample rate 45 000, has been scored 13% CER; this is the outperforming model result. This model has been compared with the baselines. As Table 1 shows our model has outperformed the baseline models; Google Lens scored 23% and Tesseract base model scored 67.8% CER.

Table 1. Models Performance Results on the Test Data

Models	CER
Propose Model	13%
Google Lens	23%
Tesseract base-model	76.8%

A sample text has been taken to show the performance difference. The sample text has been running on the three models. As Fig. 5 shows the proposed model has been outperforming the baselines. Our model exhibited a noteworthy performance in transcribing typewritten documents, achieving a commendable Character Error Rate (CER) of 4%. In stark contrast, Google Lens demonstrated a substantially higher CER at 23% and Tesseract CER at 41.5%, underscoring a considerable disparity in accuracy. It is important to note that drawing definitive conclusions based on a singular test may be premature. Therefore, to ensure the robustness of this study's findings, comprehensive testing has been carried out on a dataset comprising 20 typewritten documents. In each instance, our model consistently outperformed Google Lens, with an accuracy gap ranging from 5% to 20%, and the Tesseract by a large gap above 20%. This performance result is indicative of a promising start as we continue to advance and refine our model.

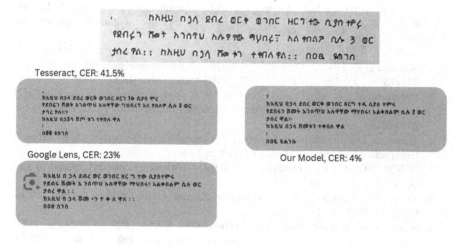

Fig. 5. Sample Transcription result

6 Conclusion

In the dynamic tech landscape, innovation remains essential, particularly in regions like Africa. OCR technology, much like the advent of writing, stands as a critical pillar. Our study, focusing on typewritten Amharic text recognition, holds promise for preserving Ethiopian archives and extending information access to the visually impaired, considering their substantial numbers in Ethiopia.

Tailoring an OCR model for Ethiopic scripts represents a significant advancement, as affirmed by our analysis and CER evaluation, which showcase the effectiveness of our model and its utilization of the Tesseract OCR engine's capabilities. The open-source nature and accuracy of Tesseract make it an ideal choice. While the challenge of limited training data for Ethiopic scripts requires creative solutions, our OCR model fills a vital niche in AI language technology, enhancing transcription accuracy and extending the reach of AI-driven OCR technology to local languages, promoting inclusively. In our study, the model performance scored 13% CER from the test data, which is 87% accuracy. Compared with the baselines that are the Tesseract base model and Google Lens platform, our model has outperformed in transcribing the Ethiopic typewritten script image or scanned documents to a digital editable format.

Looking ahead, potential improvements, such as post-processing techniques such as spell correction algorithms, can enhance our model's capabilities. In conclusion, our study underscores the transformative power of technology, and its capacity to enrich lives, preserve cultural heritage, and surmount barriers, while emphasizing the importance of ongoing innovation in addressing contemporary challenges for a brighter future.

References

1. Gebremichael, H.T., Mengistu, T.M., Beyene, M.M., Mengistu, F.G.: OCR system for the recognition of ethiopic real-life documents. In: Berihun, M.L. (ed.) ICAST 2021. LNICST, vol. 411, pp. 559–574. Springer, Cham (2022). https://doi.org/10.1007/978-3-030-93709-6_38
2. Teshome, A.: Recognition of Amharic Braille. Research gate (2009). https://doi.org/10.13140/RG.2.1.1306.3284, https://www.researchgate.net/publication/303773888_RECOGNITION_OF_AMHARIC_BRAILLE
3. Rahmati, M., et al.: Printed persian OCR system using deep learning. IET Image Process. 14(ch115), 3920–31 (2020). https://doi.org/10.1049/iet-ipr.2019.0728
4. Tesseract Documentation: Tesseract documentation | Tesseract OCR. https://tesseract-ocr.github.io/. Accessed 31 Oct 2023
5. i2OCR: i2OCR - Free Online OCR. https://www.i2ocr.com/. Accessed 31 Oct 2023
6. Vijayarani, S., Sakila, A.: Performance comparison of OCR tools. Int. J. Ubi-Comp (IJU) 6(3), 19–30 (2015)
7. Shapovalov, Y.B., Zhanna, I.B., Artem, I.A., Viktor, B.S., Aleksandr, D.U.: The potential of using Google expeditions and google lens tools under STEM-education in Ukraine. arXiv preprint arXiv:1808.06465 (2018)

8. Abish, B., Chhetri, G.B., Bhattarai, K., Pandey, M.: Nepali OCR. (2023)
9. Easy OCR Converter: Free Online OCR in 100+ Languages. Free Convert Image to Text Online. https://www.easyocrconverter.com/. Accessed 31 Oct 2023
10. Cowell, J., Hussain, F.: Amharic character recognition using a fast signature based algorithm. In: Proceedings on Seventh International Conference on Information Visualization, 2003. IV 2003, pp. 384–389. IEEE (2003)
11. Assabie, Y., Bigun, J.: Lexicon-based offline recognition of amharic words in unconstrained handwritten text. In: 2008 19th International Conference on Pattern Recognition. IEEE (2008). https://doi.org/10.1109/ICPR.2008.4761145
12. Meshesha, M., Jawahar, C.V.: Optical character recognition of Amharic documents. Afr. J. Inf. Commun. Technol. **3**(2) (2007)
13. Belay, B., Habtegebrial, T., Liwicki, M., Belay, G., Stricker, D.: Factored convolutional neural network for Amharic character image recognition. In: 2019 IEEE International Conference on Image Processing (ICIP), pp. 2906–2910. IEEE (2019)
14. Addis, D., Liu, C. M., Ta, V.D.: Printed ethiopic script recognition by using LSTM networks. In: 2018 International Conference on System Science and Engineering (ICSSE), pp. 1–6. IEEE (2018)
15. Samuel, M., Schmidt-Thieme, L., Sharma, D.P., Sinamo, A., Bruck, A.: Offline handwritten amharic character recognition using few-shot learning. In: Girma Debelee, T., Ibenthal, A., Schwenker, F. (eds.) Pan-African Conference on Artificial Intelligence. PanAfriCon AI 2022. Communications in Computer and Information Science, vol. 1800, pp. 233–244. Springer, Cham (2022). https://doi.org/10.1007/978-3-031-31327-1_13
16. Abdurahman, F., Sisay, E., Fante, K.A.: AHWR-Net: offline handwritten amharic word recognition using convolutional recurrent neural network. SN Appl. Sci. **3**, 1–11 (2021)
17. Gondere, M.S., Schmidt-Thieme, L., Sharma, D.P., Scholz, R.: Multi-script handwritten digit recognition using multi-task learning. J. Intell. Fuzzy Syst. **43**(1), 355–364 (2022)
18. Malhotra, R., Addis, M.T.: End-to-end historical handwritten ethiopic text recognition using deep learning. IEEE Access **11**, 99535–99545 (2023). https://doi.org/10.1109/ACCESS.2023.3314334

Compressed Amharic Text: A Prediction by Partial Match Context-Modeling Algorithm

Yalemsew Abate Tefera[1]($\boxtimes$)(iD), Tsegamlak Terefe Debella[2](iD),
Habib Mohammed Hussien[1](iD), and Dereje Hailemariam Woldegebreal[2](iD)

[1] Department of Electrical and Computer Engineering, Artificial Intelligence and Robotics Center of Excellence, Addis Ababa Science and Technology University, Addis Ababa, Ethiopia
yalemsew.abate@aastu.edu.et
[2] School of Electrical and Computer Engineering, Addis Ababa Institute of Technology, Addis Ababa University, Addis Ababa, Ethiopia

Abstract. Amharic is one of the most widely spoken and written languages in Ethiopia, establishing a growing presence in the digital realm. The language serves as the official working language in Ethiopia, employing its distinctive writing system, Fidel, which descends from Ge'ez characters. Amharic symbols are represented by 16 bits in Universal Transformation Format (UTF-8), a standard encoding scheme that ensures compatibility across different platforms and devices. However, this representation utilizes more bits than necessary, as a prior study demonstrated that the entropy of written Amharic language can be as low as 1.074 bits/symbol if the first-order and higher-order statistical dependencies between successive symbols in the language are well-captured. The entropy provides a lower bound on the average number of bits needed to represent a symbol without loss of information, directly impacting the achievable compression rate. This paper proposes using the Prediction by Partial Match (PPM) context-modeling algorithm to efficiently compress Amharic text. PPM is well-suited for textual data, and it adaptively uses a combination of Markov models of different orders to capture first-order and higher-order statistical dependencies between symbols in a text. We used two versions of the PPM algorithm, PPMC and PPMD, to encode eight Amharic text files. Our results show that the best order for efficient encoding is order-3. With this order, we achieved an average of 84.2% reduction in file size, which has a compression rate of 3.3 bits/symbol. This compression rate closely approximates the estimated entropy of the Amharic language (1.074 bits/symbol), demonstrating that PPM effectively compresses Amharic text by exploiting the language's inherent statistical structure. The reduced bitrate translates to significant bandwidth and energy savings in digital communication systems, as well as reduced storage space requirements.

Keywords: Amharic Language · Context-Modeling · PPM · Compression

T. G. Debelee et al. (Eds.): PanAfriConAI 2023, CCIS 2068, pp. 262–279, 2024.
https://doi.org/10.1007/978-3-031-57624-9_15

1   Introduction

Before the advent of digital communication and the information age, people sought to optimize communication with written text messages in a variety of ways. The codebook by the British Admiralty (1790), Braille (1820), and Morse code (1838) are some of the earliest examples [3]. These methods of text compression were developed to save transmission bandwidth, reduce communication time, and improve human-machine interface [3]. Despite significant advances in communication speed and availing transmission bandwidth since the early 19th century, the goal of text encoding remains unchanged: to reduce the amount of data that must be transmitted or stored. With today's proliferation of communication platforms, a significant amount of textual data is generated, transmitted, and stored, making data compression essential for efficient communication and storage.

Source encoding, sometimes called compression, is the first step in digital communications, representing outputs from an information source (e.g., voice, text, or video) in binary or digital form by removing unnecessary or redundant information from the source. This process of removing redundancy reduces the amount of data that needs to be transmitted or stored. Most written textual data comes from natural languages. Therefore, efficient text encoding relies on extracting redundancies inherent in the structure of these languages. Knowledge of the entropy of a given source language and a suitable source encoding algorithm are essential for removing the redundancies inherent in the language. Source entropy measures the average amount of information per generated symbol of the source language [7,9].

Source encoding in communication systems exploits the uncertainty associated with symbols generated by the source language to improve the efficiency and reliability of data transmission [3,9]. Despite the uncertainty, useful (meaningful) information has a discernible order, i.e., interdependence among successive output letters, due to various factors, such as grammatical rules in natural language texts. Consequently, by carefully analyzing such order, some part of the information generated by the source becomes predictable given there is a knowledge of the previously generated symbols. In reality, the predictable portion of the information can be considered redundant since it can be regenerated at the receiving end of the communication system by transmitting only the essential parts to disambiguate prediction [3,9].

Source encoding typically consists of two steps: *modeling* and *coding*. Modeling represents redundancies in a given language using a mathematical model that produces a set of probabilities based on the relationship of successive symbols [9]. The second step called *coding*, maps these probabilities to a binary code, i.e., a string of 1's and 0's. In practical applications, these two steps are employed to get close to compression ratio (coding rate) determined by the entropy of the source. However, to approach the limit set by the source's entropy, it is necessary to employ a model that captures dependencies between symbols. Moreover, the model should be compatible with a coding scheme that optimally utilizes the available code space [3,9]. In practice, a combination of *context modeling* techniques and efficient coding algorithms such as *Arithmetic coding* often serve this purpose. A context model, also known as a Finite-Context model, is a probability model that uses a few preceding symbols, the context, to estimate the probability of the next symbol [3]. These probabilities determine the number of bits required to code the symbol. The model construction is achieved incrementally by calculating probability estimates from frequency counts of symbols in the text source.

A pioneer work done on entropy estimation and entropy-based encoding of the Amharic language, which is the focus of this paper, was presented in [11]. In the paper, Shannon's N-gram model was used to estimate block entropy (G_N) and conditional entropy (F_N), up to $N = 60$, using corpora of six books, all written in Amharic language, spanning different domains. Based on the reported results, the language's minimum number of bits per symbol (zero crossing entropy) H_Z was found to be 1.074-bits/symbol at $N = 15$ with a 1% error margin [11]. Furthermore, the paper utilized entropy coders, namely Huffman coding and Arithmetic coding, to see how practical source coders approach the estimated entropy. In this regard, the paper identified that the language has an average code rate of 5.805-bits/symbol and producing a 72.45% reduction in file size for first-order language model (at $N = 1$) [11]. Overall, the compression result agrees with the estimated entropy of the language, and the paper concluded that compression can be improved beyond the theoretical estimations (86.54% redundancy) since Amharic symbols are currently allocated a fixed 16-bit/symbol in the UTF-8 encoding standard [11].

With this observation in mind, our paper proposes to further investigate the claims made in [11] using the prominent context-modeling algorithm called *Prediction by Partial Match (PPM)*. We propose PPM to demonstrate the advantages of context-modeling for encoding Amharic text sources and to quantify how close practical compression gets to the estimated entropy of the language. We also determine the optimal order for efficient compression of Amharic text sources. The Prediction by Par1111111tial Match (PPM) algorithm is known for its ability to achieve high compression rates. It has been extensively applied in compression studies involving various languages, including English [4,5], Arabic, Armenian, Chinese, Persian, and Russian [1]. Table 1 provides a summary of the compression results for these languages, measured in *bits per symbol*.

Table 1. PPM compression results for different languages [1].

Language	Corpora	Order						
		1	2	3	4	5	6	7
Arabic	BAAC	2.42	2.17	2.04	1.79	1.66	1.51	1.44
Armenian	HC	2.43	2.02	1.90	1.69	1.61	1.42	1.36
Chinese	LCMC	4.03	3.01	2.66	2.49	2.46	2.47	2.49
English	Brown	3.67	3.02	2.51	2.23	2.16	2.17	2.22
	LOB	3.66	2.99	2.48	2.21	2.14	2.15	2.19
Persian	Hamshahri	2.29	2.09	1.96	1.75	1.6	1.42	1.33
Russian	HC	2.47	2.08	1.97	1.73	1.63	1.41	1.33

2 Information Source and Lossless Source Encoding

In practice, a digital communication system is often modeled as shown in Fig. 1 [7]. In the model, the information source can be either discrete in time with a finite (countable) alphabet, such as written languages, or analog (continuous in time with an uncountable alphabet) such as audio or video. However, despite the difference in the nature of an information source, in a digital communication system, messages produced by the source are transformed into a sequence of binary digits by the source encoder.

Ideally, source encoding should represent the source's output symbols or messages with the fewest possible binary digits. To achieve this goal, an efficient

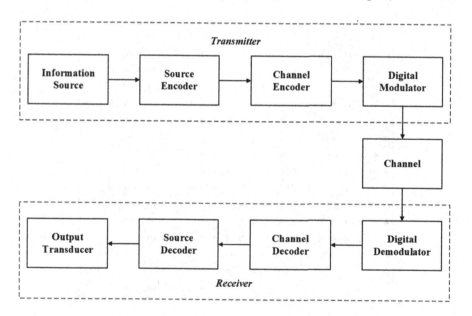

Fig. 1. Basic Components of a Digital Communication System [7].

representation of the source output that minimizes redundancy is required. In this aspect, the source encoder ensures efficient utilization of the channel by compressing the original message to an optimal size.

2.1   The Information Source

Amharic (አማርኛ) is a Semitic language predominantly spoken in Ethiopia. According to the 1994 Ethiopian census conducted by Ethnologue [13], ninety languages are spoken in Ethiopia. Among these languages, Amharic is one of the official languages of the Federal Government and is the most popularly utilized in various regions of the country. In addition, Amharic is the second most widely spoken Semitic language after Arabic [15].

The writing system of Amharic has a unique set of semi-syllabic symbols called Fidel (ፊደል) [13]. The system has a total of 182 basic symbols, comprising 26 consonant symbols each followed by vowel coupled seven variations. In addition to these basic symbols, the Amharic Fidel has an additional 56 symbols which are mainly used to account for inter-indigenous language translation, 40 ligatures, 9 punctuation marks, 20 number symbols, and 3 combining marks. In digital applications, the current encoding of Amharic in UTF-8 (the dominant version of Unicode) allocates a 16-bit code range (1200-137F) also known as the Ethiopic Unicode block. Figure 2 shows the symbols included in the Ethiopic Unicode block, which is composed of symbols found in writing Amharic, G-e'ez (ግእዝ), Tigrinya (ትግርኛ) and other Ethiosemitic languages [12]. Overall, in the UTF-8 encoding, the language is allocated a total of 384 code points, including additional 8 punctuation marks and 26 reserved code points [12].

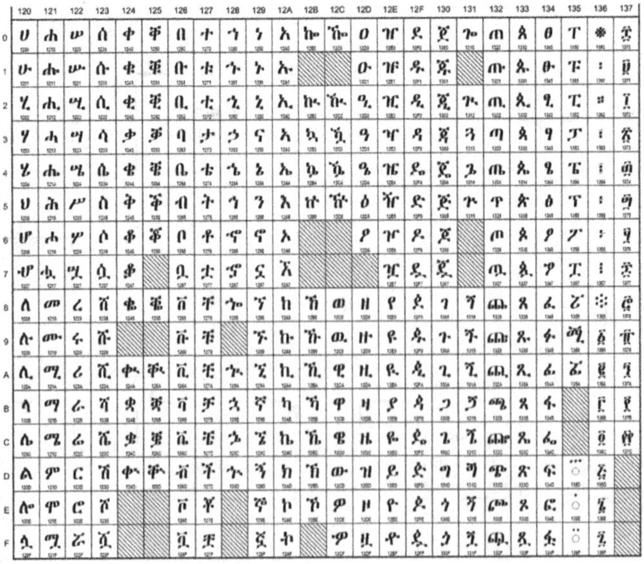

Fig. 2. Ethiopic Unicode Table [13].

2.2   Lossless and Lossy Source Encoding

Based on the requirements of reconstruction, source encoding schemes can be divided into two classes: *lossless* source encoding and *lossy* source encoding [9]. As the name implies, in lossless source coding there is no loss of information when the message is reconstructed back from the binary sequences that are obtained from a source coder; i.e., the coding is completely reversible. Lossless encoding is preferred for encoding texts in written languages because it preserves all of the information in the original text. This is important because even a change in one symbol can potentially alter the entire meaning of the conveyed information.

Lossy source coding schemes, on the other hand, involve the loss of some information during compression and exact reproduction of the original message is not possible. Such source encoding techniques often make sense for compressing image, audio, or video sources. For such information sources, if the loss of data is handled carefully, it may not make a noticeable difference during reconstruction as the human ear and eye are capable of smoothing (filtering) out minor losses [6].

2.3   Lossless Source Encoding Performance Measures

The following measures are commonly used to express the performance of source encoding (compression) algorithms [9]:

Compression Ratio. The compression ratio (CR) is the ratio of the number of bits required to represent data before compression to the number of bits required to represent the data after compression.

$$CR = \frac{Uncompressed\,Size}{Compressed\,Size} \tag{1}$$

Space Savings. The average amount of space saved after compression is given by:

$$Space\,Savings = \left(1 - \frac{1}{CR}\right) \times 100\% \tag{2}$$

Compression Rate. The compression rate is the average number of bits required to represent a certain sample. Its measure is always dependent on how the sample units are represented. For example, for text sources, the rate is the average number of bits required per symbol (bits/symbol) and for image sources, it is given in terms of pixels (bits/pixel).

$$Compression\,Rate = \frac{Compressed\,Size}{Number\,of\,Sample\,Units} \tag{3}$$

Compression rate serves as a valuable metric for assessing the efficiency of source encoding algorithms within the context of communication systems. It quantifies the level of compression achieved in relation to the original size of the data. In communication systems, a higher compression rate indicates more efficient utilization of bandwidth or transmission resources.

However, when evaluating performance in storage applications, a different performance metric is commonly employed. In this context, the compression ratio or space savings are typically presented as the primary effectiveness indicators. The compression ratio represents the ratio of the original data size to the compressed data size, providing insights into the level of data reduction achieved. Space savings, on the other hand, refer to the amount of storage space saved through compression, which is particularly relevant in storage-constrained environments.

3   Modeling of Text Sources

Modeling in source encoding, in general, is the process of identifying and extracting redundancy from the information source and generating an efficient representation that only captures the non-redundant information [9]. Similarly, modeling text sources is essentially modeling of natural languages, which are sources of written text. Observing the patterns exhibited in different languages, redundancy can be attributed to two primary reasons: the unequal probability of occurrence of symbols and the interdependence of symbols resulting from grammatical rules, commonly referred to as context.

Broadly speaking, source models for written languages are often classified as Morphological and Statistical [15]. Morphological methods rely on grammars which are defined based on linguistic knowledge [15]. On the contrary, statistical language models (SLMs) use a corpus to generate probability models based on the frequency of occurrence of symbols and the correlation between symbols [11, 15]. Consequently, SLMs could assign probabilities to symbols (group of symbols) using one of the following two approaches [8]:

1. Frequency: symbol probabilities are calculated by counting their frequencies of occurrences;
2. Context: the preceding symbols (the context) is considered

When employing the second approach for practical coding applications, the maximum length of the context is usually limited to a fixed length to ensure synchronization between the encoder and the decoder.

In order to study the statistical properties of a language through SLMs a collection of texts written in that language (corpus) has been widely used [1,3,11]. In this regard, Claude Shannon modeled written English texts as a random process, where a source generates a sequence of symbols from a finite alphabet with unequal probabilities [10]. This random process is assumed to be ergodic and the probability of occurrence of the next symbol in the text depends on the previous symbols (or its context). In practice, this random process is approximated by a Markovian process [9,10]. The Markov process has been utilized to model a range of natural languages including two Ethiopian local languages, i.e., the

Amharic and Afan Oromo [11,14]. Consequently, for this paper, we assume the Amharic language to follow the properties of a Markovian process.

Markovian models are used to represent statistical dependence between sequences of symbols on each other [9]. For models used in lossless source encoding of discrete sources, a specific type of Markov process, called discrete-time Markovian chain is used. In this model, knowledge of the past k symbols is presumed to be sufficient to predict the next symbol. Overall, if we let $\{x_n\} = \{x_1, x_2, \ldots, x_n\}$ to be a sequence of observed symbols, a kth-order Markov model of the given sequence is given as [9]:

$$P(x_n|x_{n-1}, \ldots, x_{n-k}) = P(x_n|x_{n-1}, \ldots, x_{n-k}, \ldots) \tag{4}$$

Particularly, Markov models become important to model text sources since the probability of occurrence of a symbol is heavily influenced by the preceding letters due to the presence of grammatical structures in natural languages. Moreover, in most source encoding algorithms, knowing the context of a symbol being encoded often makes it possible to obtain an accurate probability model for a symbol. However, the idea of using longer contexts to determine the probability model of a symbol often quickly becomes complex since the number of permutations of contexts for a symbol grows exponentially with the size of the context [3,9]. To better understand the extent of this problem, consider an alphabet of size N; then, the number of first-order contexts, i.e., if we assume each symbol to be independent, is N. However, if we assume each symbol is dependent on its immediate predecessor, the number of second-order contexts is N^2. Thus, to encode a sequence from an alphabet of size 200 using contexts of order-5, $200^5 = 3.2 \times 10^{11}$ probability distributions need to be calculated. In reality, this is not often a practical approach since it requires a huge set of data and becomes computationally exhaustive. Consequently, almost all practical context modeling approaches are adaptive. Using the adaptive approach the probability distributions are generated while the text is being encoded. Thus the statistics grow proportionally with the length of the text processed.

Adaptive context models rely on the context provided by preceding symbols to predict subsequent symbols. In the previous statement, the term "predict" implies that the frequency counts that determine the encoding of the current character are associated with its context. While encoding the symbol, the frequency distributions are used to calculate symbol probabilities which in turn determine the number of bits the symbol contributes to the binary representation. Generally, there are two ways in which a context model uses previously occurred characters. It may use a fixed number of previous characters in its predictions or combine predictions from contexts of several lengths [3].

1. Fixed order model: A model that uses k previous characters to predict the current character is called an order-k fixed-context model. k can take values starting from 0, where the context is not considered and coding simply takes one character at a time. However, when $k = 1$, the preceding charter is used in encoding the current character. Similarly if $k = 2$, the previous two characters become the context and continue in this manner for higher-order models.

2. Blended model: Blending combines models with different values of k to improve symbol probability calculations. For example, if the maximum context length is set to 3, i.e. the model uses the previous three characters, when the three-character context fails to predict, the model tries to predict using a reduced order-2 context model. In the case where all the available models (orders 3, 2, 1, and 0) fail to encode the symbol, an order-(-1) model will be used. The order-(-1) model is the default fallback model where each symbol in the alphabet is equiprobable. In practice, symbol probability calculations in a blended model can span all sub-models, i.e. the maximum-length context and all lower-order contexts. This type of model is known as a *fully-blended model*. Alternatively, a *partially-blended model* considers only some of the lower-order context models [3].

4 PPM Context-Modeling

PPM is a compression method first proposed by John Cleary and Ian Witten [4], with extensions and implementation by Alistair Moffat [5]. It uses a partially blended finite-context model in combination with arithmetic coding. Not only it can be a good approximation to the entropy but also helps to incrementally improve coding which makes it a good fit with arithmetic coding.

Context models try to capture the correlation between successive symbols and provide an approximate model of text sources. By utilizing higher-order Markov models, an improved compression ratio can be achieved. However, using such models is impossible at the initial phase of coding because enough statistics is not available to build higher-order models. Due to this factor, coding becomes relatively inefficient at the start of the encoding process. This problem is common to adaptive coding schemes, as models are constructed incrementally based on the text source itself. To address this issue, PPM introduces a compromise, a strategy known as a "partial match" approach. This involves building a higher-order model but using it for lower-order predictions in cases when high-order ones are not yet available [3,4].

During the adaptive process of estimating symbol probabilities, especially at the beginning, symbols that have not been seen before for any context are encountered. This is called the zero frequency problem [3,9]. In order to handle this situation in PPM the source coder alphabet contains a special "escape (esc)" symbol, which is used to signal that the symbol to be encoded has not been seen in the current context [3,4].

By combining the partial match strategy and the escape symbol mechanism, PPM implements a partial blending scheme where the model is based on variable length contexts. The algorithm initially attempts to use the longest available context to predict the next symbol. When the symbol to be encoded has not previously been encountered in the current context, an escape symbol is encoded and the PPM encoder shortens the context by dropping one symbol. This process continues until a context that has a probability estimate of the symbol is found. In cases where an estimate is not found in any sub-contexts of the current context a default model is used to encode each symbol with probability $1/M$ where M is the size of the source alphabet.

4.1   Escape Symbol Probability

While assigning probability to symbols in a certain context PPM reserves some probability space and assigns it to the escape symbol [3,9]. This approach is used to handle situations where novel symbols are encountered, ones that have not been previously observed within the given context. It also helps to continue encoding since arithmetic coding requires all symbols to have a non-zero probability. In this way, the PPM decoder can follow the encoder while looking for a symbol in the perspective context. When the encoder decides to switch to a shorter context, it first encodes the escape symbol. Accordingly, the decoder will be able to decode the escape symbol, since it is encoded in the current context. Up on decoding the escape symbol, the decoder also switches to the shorter context, ensuring synchronization with the encoder.

With the above discussion in mind and assuming the maximum length of a context is K, for each context C of length $k \leq K$, PPM allocates a probability $P_k(esc|C)$ for all symbols that did not appear after the context C and the remaining $1 - P_k(esc|C)$ is distributed among all other symbols that have non-zero counts with C.

4.2   PPM Symbol Probabilities

Let Ω be a text source with a discrete alphabet,

$$\mathcal{A}_\Omega = \{a_1, a_2, \ldots, a_M\} \tag{5}$$

consisting M symbols with $M \geq 2$.
Define:

α^n : a sequence of n symbols taken from some part (sample) of the source (text);

$$\alpha^n = \{\alpha_1\alpha_2\ldots\alpha_n\} : \alpha_i = a_m$$
$$m = 1, 2, \ldots, M \ \& \ i = 1, 2, \ldots, n \tag{6}$$

K : the order of the longest context (maximum order of a Markov model)
C_k : context with length $k \leq K$

Let's assume the first $j - 1$ symbols $\alpha_1\alpha_2\ldots\alpha_{j-1}$ of the sequence α^n are already processed
The probability of the next symbol α_j in the sequence α^n with respect to context C_k,

$$C_k = \alpha_{j-k}\alpha_{j-k+1}\ldots\alpha_{j-2}\alpha_{j-1} \tag{7}$$

which contains the previous k symbols before α_j, is [2],

$$P(\alpha_j|C_k) = \begin{cases} P_k(\alpha_j|C_k) & \text{if } C_k\alpha_j \text{ appeared before,} \\ P_k(esc|C_k)P(\alpha_j|C_{k-1}) & \text{Otherwise.} \end{cases} \quad (8)$$

Where:

$$P_k(\alpha_j|C_k) = \frac{\mathbb{C}_{C_k}(\alpha_j)}{\mathbb{T}_{C_k} + \mathbb{C}_{C_k}(esc)} \quad (9)$$

$$P_k(esc|C_k) = \frac{\mathbb{C}_{C_k}(esc)}{\mathbb{T}_{C_k} + \mathbb{C}_{C_k}(esc)} \quad (10)$$

$$\mathbb{T}_{C_k} = \sum_{a_m \in \mathcal{A}_{C_k}} \mathbb{C}_{C_k}(a_m) \quad (11)$$

$$\mathcal{A}_{C_k} = \{a_m : \mathbb{C}_{C_k}(a_m) > 0\}, \ m = 1, 2, \ldots, M \quad (12)$$

$P_k(\alpha_j|C_k)$: symbol α_j's probability in context C_k
$P_k(esc|C_k)$: escape probability in context C_k
$\mathbb{C}_{C_k}(\alpha_j)$: count of symbol α_j in context C_k (number of times $C_k\alpha_j$ appeared in the text)
$\mathbb{C}_{C_k}(esc)$: count of the escape symbol in context C_k
$\mathbb{T}_{C_k}$: total count of symbols in context C_k
$\mathcal{A}_{C_k}$: alphabet containing distinct symbols that followed context C_k
C_{k-1} : next smaller sub-context of C_k, length $k-1$

PPM encoder uses frequency counts of symbols ($\mathbb{C}_{C_k}$) in each context C_k to provide cumulative probability to an arithmetic encoder. As an example, Table 2 shows the assignment of symbol probabilities $P_k(\alpha_j|C_k)$ for the phrase "ለምለም ፉ.ለፉ.ለፕ" (particularly chosen because of repetition of symbols). The space between words is represented by "⌣". The longest context is assumed to use the previous two symbols ($K = 2$). The escape symbol count is assigned using a method called PPMC; i.e. the number of distinct symbols that appeared in the context.

4.3   PPM Variants

There are four variants of PPM based on the assignment of escape counts. There is no theoretical basis for choosing the escape probability optimally [3]. The methods have been selected based on the vast experience that developers had with data compression.

PPMA. In Method A the escape symbol count is set to 1 in each context ($\mathbb{C}_{C_k}(esc) = 1$). This gives the escape probability of [3].

$$P_k(esc|C_k) = \frac{1}{\mathbb{T}_{C_k} + 1} \quad (13)$$

The probability assigned to symbol α_j (different from the escape symbol) in context C_k becomes [3].

$$P_k(\alpha_j|C_k) = \frac{\mathbb{C}_{C_k}(\alpha_j)}{\mathbb{T}_{C_k} + 1} \quad (14)$$

Table 2. PPM model for string ለምለም፡ ፈለፈለች

(a) Order-2

Order $k = 2$			
Context	Symbol	$\mathbb{C}_{C_2}$	P_2
ለም	ለ	1	1/4
	‿	1	1/4
	esc	2	2/4
ምለ	ም	1	1/2
	esc	1	1/2
ም‿	ፈ	1	1/2
	esc	1	1/2
‿ፈ	ለ	1	1/2
	esc	1	1/2
ፈለ	ፈ	1	1/4
	ች	1	1/4
	esc	2	2/4
ለፈ	ለ	1	1/2
	esc	1	1/2

(b) Order-1

Order $k = 1$			
Context	Symbol	$\mathbb{C}_{C_1}$	P_1
ለ	ም	2	2/7
	ፈ	1	1/7
	ች	1	1/7
	esc	3	3/7
ም	ለ	1	1/4
	‿	1	1/4
	esc	2	2/4
‿	ፈ	1	1/2
	esc	1	1/2
ፈ	ለ	2	2/3
	esc	1	1/3

(c) Order-0

Order $k = 0$		
Context	Symbol	$\mathbb{C}_{C_0}$	P_0
	ለ	4	4/15
	ም	2	2/15
	‿	1	1/15
	ፈ	2	2/15
	ች	1	1/15
	esc	5	5/15

PPMB. Method B treats the appearance of a new symbol in a context as a rare event when a few symbol(s) followed a particular context beforehand [3]. In practice, the encoding of a symbol is delayed until it appears more than once in the context. This is done by subtracting one from all counts and giving the count to the escape symbol.

Let n_{C_k} be the number of distinct symbols that have occurred in context C_k. The escape probability using method B is [3],

$$P_k(esc|C_k) = \frac{n_{C_k}}{\mathbb{T}_{C_k}} \tag{15}$$

Before allowing code space for escape symbol, the probability of each symbol α_j is [3,6],

$$P_k(\alpha_j|C_k) = \frac{\mathbb{C}_{C_k}(\alpha_j) - 1}{\mathbb{T}_{C_k} - n_{C_k}} \tag{16}$$

Thus, to encode a symbol it must appear at least twice in the particular context. After reassigning the discounted values from each symbol to the escape symbol, the code space allocated for symbol α_j becomes [3,6],

$$P_k(\alpha_j|C_k) = \frac{\mathbb{C}_{C_k}(\alpha_j) - 1}{\mathbb{T}_{C_k}} \tag{17}$$

PPMC. In Method C, similar to Method B, the count assigned to the escape symbol is equal to the number of symbols that have occurred in that context $(\mathbb{C}_{C_k}(esc) = n_{C_k})$. However, instead of subtracting from the counts of individual symbols and assigning it to the escape symbol, the total count is incremented by the same amount. Consequently, the escape and symbol probabilities respectively become [3],

$$P_k(esc|C_k) = \frac{n_{C_k}}{\mathbb{T}_{C_k} + n_{C_k}} \tag{18}$$

$$P_k(\alpha_j|C_k) = \frac{\mathbb{C}_{C_k}(\alpha_j)}{\mathbb{T}_{C_k} + n_{C_k}} \tag{19}$$

PPMD. In this implementation of the PPM called method D ("D" stands for Double): symbol α_j's frequency is incremented by 2 in a certain context C_k when $\mathbb{C}_{C_k}(\alpha_j) > 0$, otherwise this 2 increment is divided between the current symbol and the escape symbol. The escape symbol gets a probability [6],

$$P_k(esc|C_k) = \frac{n_{C_k}}{2\mathbb{T}_{C_k}} \tag{20}$$

and the probability of symbol α_j is

$$P_k(\alpha_j|C_k) = \frac{2\mathbb{C}_{C_k}(\alpha_j) - 1}{2\mathbb{T}_{C_k}} \tag{21}$$

Applying these four methods, some performance variations were observed depending on the data being encoded. Overall, PPMD and PPMC provide the best compression ratio [3, 5, 6]. Incidentally, considering our objective of efficient encoding of Amharic text sources, the implementation of PPM in this paper focused on both PPMC and PPMD versions.

5   Results and Discussions

5.1   Experimental Setup

PPMC and PPMD versions of the PPM context modeling algorithm are implemented using Python. Eight books written in Amharic language are taken as test files which are presented in Table 3. It is important to note that no specific criteria were employed for selecting these books; rather, they were chosen as they were the only available dataset at the time of the study.

Table 3. Test Files

Book	Size in bytes
Tsehay ፀሀይ	154,453
Agere አገሬ	367,740
Tenbit ትንቢት	699,320
Yegazetegnaw Mastawesha የጋዜጠኛዉ ማስታወሻ	841,068
Fiker Eske Mekaber ፍቅC አስከ መቃብር	857,349
Yederasiw Mastawesha የደራሲዉ ማስታወሻ	1,033,812
New Testament አዲስ ኪዳን	1,033,812
Bible መጽሐፍ ቅዱስ	4,475,512

5.2   Average Compression Results

Average compression performance of PPMC and PPMD over the eight books for orders 0 up to 10 is presented in Tables 4 and 5:

Table 4. PPMC Average Compression Performance

Performance Metrics	Order										
	0	1	2	3	4	5	6	7	8	9	10
Compression Ratio	3.6	4.83	6.09	6.315	6.29	6.25	6.21	6.185	6.17	6.155	6.15
Space Savings	72.22	79.3	83.58	84.16	84.10	83.99	83.89	83.83	83.78	83.75	83.73
Compression Rate (bits/symbol)	5.74	4.29	3.41	3.29	3.32	3.34	3.36	3.38	3.39	3.392	3.398

Table 5. PPMD Average Compression Performance

Performance Metrics	Order										
	0	1	2	3	4	5	6	7	8	9	10
Compression Ratio	3.6	4.84	6.13	6.36	6.345	6.31	6.275	6.25	6.23	6.22	6.21
Space Savings (%)	72.22	79.33	83.67	84.27	84.23	84.14	84.06	84	83.94	83.92	83.89
Compression Rate (bits/symbol)	5.735	4.28	3.39	3.28	3.32	3.31	3.33	3.345	3.35	3.36	3.362

5.3 Discussion

Based on the results presented in Tables 4 and 5, it can be inferred that Order-3 contexts yield the highest compression ratio on average. Similarly, the plot depicted in Fig. 3 shows, that for both algorithms compression ratio rises sharply up to order-3 and falls slightly for increasing order.

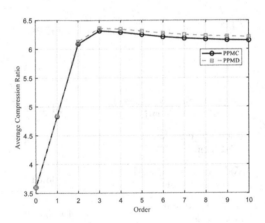

Fig. 3. Average Compression Ratio with Order for PPMC and PPMD

Thus we can conclude that order-3 is the *optimum* order for efficient compression of Amharic text sources using PPM. With increasing context length, a decrease in compression performance is expected which is also the case for other languages: for English order-4 or order-5 is taken as the optimum order [3,9]. From our observations, the reason for this slight degradation is that for relatively small orders (1 up 4) repetitions of contexts are much more likely than higher-order contexts. Consequently, using higher-order contexts requires coding the escape symbol more frequently due to the uniqueness(non-repetition) of contexts. Moreover, if the context size grows indefinitely to large orders, using the escape symbol to skip contexts becomes an overhead which makes incremental arithmetic coding inefficient. This additional overhead creates a degradation in performance by increasing the size of the compressed file.

Compared to the encoding of Amharic symbols using UTF-8, which uses 16-bits/symbol, at the optimum order-3 PPMC requires 3.29-bits/symbol (Table 4) and PPMD requires 3.28-bits/symbol (Table 5). These results show a significant gap (around 12.7-bits) which shows the advantage gained by using PPM context modeling algorithms to code Amharic text sources. Additionally, as indicated in Tables 4 and 5, an 84.2% reduction in file size is obtained by employing both algorithms. These results show around a 5% improvement compared to previous Amharic text source coding approaches presented in [11], which utilized N-gram with Huffman and arithmetic coding. The resulting space savings for N = 2 and N = 3 were 75.98% and 79.55% respectively, as reported in [11].

Another perspective to consider while examining the above results is to compare them with the theoretical entropy estimation of the Amharic language. In [11], Tesgamlak and Dereje estimated a lower bound entropy of the language to 1.074-bits/symbol. Figure 4 depicts the average compression rate for both PPMC and PPMD in bits/symbol.

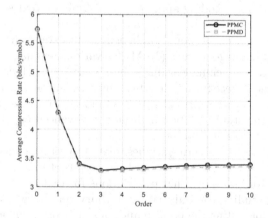

Fig. 4. Average Compression Rate with Order for PPMC and PPMD

At the optimal order-3, both algorithms require approximately 3.3-bits/symbol, which exceeds the theoretically estimated entropy of the language by 2.23 bits.

6 Conclusions

This paper aims to analyze context-modeling techniques for modeling Amharic language and coding of Amharic text sources. In this study, the PPM algorithm was selected as a primary approach, and the most effective versions of the PPM context-modeling algorithm, namely PPMC and PPMD were utilized to code Amharic text source files. In addition, the performance of these algorithms was evaluated using various metrics such as compression ratio, space savings, and average compression rate.

The experimental results demonstrated the optimum order (context length) for efficient compression (source encoding) of Amharic texts is order-3. Furthermore, for the optimal order, it was possible to achieve 84.2% reduction in file size which will significantly improve bandwidth utilization for communication applications and reduce resources required for storing Amharic text source files, as compared to UTF-8 representation.

The golden standard that all source coding algorithms aim for is to approach the entropy of the source. With this regard, the results presented in the paper have a 2.23-bits/symbol difference as compared to theoretical entropy estimations of the Amharic language. While there has been a slight improvement over previously reported source coding results in [11], the gap shows there is still room for improvement.

Finally, the findings presented in this paper exhibited the advantages that can be gained from employing context-modeling techniques for source coding applications of Amharic text sources. Mainly improvements in compression ratio and compression rate are achievable. These outcomes can provide insights into the efficient modeling and encoding approaches for Amharic text sources in various applications.

As a recommendation, it would be beneficial to extend the study by incorporating Neural Language Models (NLM) for various language processing tasks within the context of machine learning. These tasks can include speech recognition, optical character recognition, handwriting recognition, machine translation, spelling correction, image captioning, text summarization, and more. By incorporating NLMs, the study can benefit from the advanced capabilities of neural networks in language-related tasks.

Furthermore, when conducting studies related to source coding and compression, it is valuable to establish a standardized corpus of source files. This corpus typically consists of books or other publications for text sources. In the case of English language studies, the Calgary, Canterbury, and Brown corpora are commonly employed for testing the performance of source encoding and compression algorithms. Having a standardized corpus enables researchers to effectively compare and justify their results, leading to better comparability and facilitating progress in studies related to Amharic or other local languages.

References

1. Alhawiti, K.M.: Adaptive models of Arabic text. Ph.D. thesis, Bangor University, Wales (2014)
2. Begleiter, R., El-Yaniv, R., Yona, G.: On prediction using variable order Markov models. J. Artif. Intell. Res. **22**, 385–421 (2004)

3. Bell, T.C., Cleary, J.G., Witten, I.H.: Text Compression. Prentice-Hall, Hoboken (1990)

4. Cleary, J.G., Witten, I.H.: Data compression using adaptive coding and partial string matching. IEEE Trans. Commun. **32**(4), 396–402 (1984)

5. Moffat, A.: Implementing the PPM data compression scheme. IEEE Trans. Commun. **38**(11), 1917–1921 (1990)

6. Mukherjee, A., Awan, F.: Text compression. In: Sayood, K. (ed.) Lossless Compression Handbook, pp. 227–245. Academic Press, California (2003)

7. Proakis, J.G., Salehi, M.: Information source and source coding. In: Communication Systems Engineering, pp. 267–280. Prentice-Hall, New Jersey (2002)

8. Salomon, D., Motta, G.: Statistical methods. In: Handbook of Data Compression, pp. 211–227. Springer, London (2010). https://doi.org/10.1007/978-1-84882-903-9_5

9. Sayood, K.: Introduction to Data Compression. Morgan Kauffman, Massachusetts (2012)

10. Shannon, C.E.: A Mathematical Theory of Communication. Bell Syst. Tech. J. **27**(3), 379–423 (1948)

11. Terefe, T., Hailemariam, D.: Entropy estimation and entropy-based encoding of written Amharic language for efficient transmission in telecom networks. In: 2017 IEEE AFRICON, pp. 238–244. IEEE, Cape Town, South Africa (2017)

12. The Unicode Consortium: The unicode standard version 11 (2018). http://www.unicode.org/charts. Accessed 28 July 2018

13. Wimsatt, A., Wynn, R.: Amharic Language and Cultural Manual (2011). http://languagemanuals.weebly.com/language-manuals-list.html. Accessed 7 Dec 2018

14. Woldegebreal, D.H., Debella, T.T., Molla, K.D.: On the entropy of written Afan Oromo. In: Sheikh, Y.H., Rai, I.A., Bakar, A.D. (eds.) e-Infrastructure and e-Services for Developing Countries, vol. 443, pp. 25–46. Springer, Cham (2022). https://doi.org/10.1007/978-3-031-06374-9_3

15. Yifru, M.: Morphology based-language modeling of Amharic. Ph.D. thesis, Hamburg University, Germany (2010)

Author Index

T. G. Debelee et al. (Eds.): PanAfriConAI 2023, CCIS 2068, pp. 281–282, 2024.
https://doi.org/10.1007/978-3-031-57624-9

Printed in the United States
by Baker & Taylor Publisher Services